THE RISE AND FALL OF THE NEW CHRISTIAN RIGHT

THE RISE AND FALL OF THE NEW CHRISTIAN RIGHT

Conservative Protestant Politics in America 1978–1988

STEVE BRUCE

CLARENDON PRESS · OXFORD

Oxford University Press, Walton Street, Oxford OX2 6DP
Oxford New York Toronto
Delhi Bombay Calcutta Madras Karachi
Petaling Jaya Singapore Hong Kong Tokyo
Nairobi Dar es Salaam Cape Town
Melbourne Auckland
and associated companies in
Berlin Ibadan

Oxford is a trade mark of Oxford University Press

Published in the United States
by Oxford University Press, New York

First published 1988
First published in Clarendon Paperbacks 1990
Paperback reprinted 1992

British Library Cataloguing in Publication Data
Bruce, Steve
The rise and fall of the new Christian right: conservative protestant politics in America 1978–1988
1. United States. Christian right-wing political movements
I. Title
322'.0973
ISBN 0–19–827861–6

Library of Congress Cataloging in Publication Data
Bruce, Steve, 1954–
The rise and fall of the new Christian right: conservative Protestant politics in America, 1978–1988/Steve Bruce.
p. cm.
Bibliography: p.
Includes index.
1. Evangelicalism—United States—History—20th century. 2. Fundamentalism—History—20th century. 3. Conservatism—United States—History—20th century. 4. Christianity and politics—History—20th century. 5. United States—Politics and government—1977–1981. 6. United States—Politics and government—1981– 7. United States—Church history—20th century
I. Title.
BR1642. U5B78 1988 320.5'5'0973—dc19 88–19786
ISBN 0–19–827861–6

Printed and bound in
Great Britain by Biddles Ltd,
Guildford and King's Lynn

This book is dedicated to Peter Metcalf, whose friendship, veranda, and hideous household gods made visits to Virginia so pleasant

PREFACE TO THE PAPERBACK EDITION

A PROBLEM of topical writing is the constant need to update. In December 1987 I added the Afterword to report on Jerry Falwell's resignation from active leadership of the Moral Majority and Liberty Federation. Now, two years later, there are many other observations I could make about the fate of socio-moral politics in America. Fortunately for my claims to prescience, the career of socio-moral politics over the intervening years has proceeded in much the way one would expect from my general argument that American fundamentalists are (*a*) sufficiently numerous that when organized they can bring their issues into the public arena for discussion but (*b*) too weak to significantly change the socio-moral climate of America.

For analysts, the most important event of 1988 was the Reverend Pat Robertson's campaign to win the Republican party nomination for the presidency. Despite his attempts to alter his biography from that of a television evangelist to that of a successful conservative businessman whose business just happened to be religious television (see pp. 129–31), Robertson was seen by both his supporters and his detractors as a standard-bearer for the new Christian right (NCR). Both his primary election results and the vast body of opinion polling that accompanies such campaigns give us considerable insight into public responses to the NCR, and what Robertson's campaign showed was that, contrary to the inflation of many journalists and some social scientists, the more exposure the NCR had, the less popular it became.[1]

In September 1987, a *Time* poll asked Republican voters and 'leaners' who they 'would be proud to have as President', and they chose in the following proportions:

	%
George Bush	69
Bob Dole	68
Jack Kemp	58
Pete Du Pont	49
Al Haig	46
Pat Robertson	26

[1] The points made in this preface are discussed at length in S. Bruce, *Pray TV: televangelism in America* (Routledge, London, 1990), where one will find a detailed evaluation of Robertson's campaign and bibliographical information on the sources for the claims made here.

Not surprisingly, Robertson also came last in questions which asked about political experience or 'ability to deal with the Soviet Union', but when asked 'is —— someone you can trust?', respondents put Robertson last, behind four professional politicians and a soldier. Even more significant is the distance between the most and the least trusted. George Bush came first with 80 per cent. Robertson was trusted by only 43 per cent of those asked.

Even in the South, with its concentration of Baptist fundamentalists, Robertson was not popular. A Roper poll showed that only 16 per cent of adults in twelve southern states said they would consider voting for Robertson. More importantly as a sign of the hostility he provoked, 69 per cent said they felt negative towards his campaign, the worst negative rating of all twenty potential Republican and Democratic candidates.

Once the primaries got under way Robertson's chances seemed to improve, but this was an artefact of the nominating system. States differ in the ways they organize their primaries. In some there is an election open to all registered Republicans; in others there is a caucus which involves people actually turning up at halls and being counted. In small states and in those which used the caucus system the high degree of commitment of Robertson's following disguised its smallness. But once the campaigns moved on to the large states with open primary elections, Robertson slipped badly. 'Super Tuesday'—the simultaneous primaries across the South—finished him. A poll in late February 1988 showed only 15 per cent likely to support him and, with 72 per cent saying they 'would definitely not vote for' him, Robertson was less popular overall than even the most liberal Democrat. The most significant poll finding concerned the intentions of self-identified evangelicals and fundamentalists among Republican voters. They divided 44 per cent for Bush, 30 per cent for Dole, and only 14 per cent for Robertson. When asked if his former status as a clergyman made them more or less likely to vote for him, even conservative Protestants answered 'less likely', in proportion of 42 to 25 per cent.

In the event the polls were remarkably accurate. In almost every state, including his home state of Virginia, Robertson finished a poor third and was beaten by Bush in every demographic group including fundamentalists.

This outcome is entirely consistent with the arguments of this book. There were particular circumstances which did not help

Robertson. The Bakker sex scandal (pp. 143–5) was followed by the disgrace of leading television evangelist Jimmy Swaggart (also for sexual misdemeanours), and the two together seriously undermined the public standing of religious broadcasters. The damaging fall-out continued through 1988 and into 1989 with Jim Bakker being convicted of fraud and sentenced to thirty years' imprisonment and Jimmy Swaggart rejecting the discipline of his denomination, the Assemblies of God. However, the largest part of the explanation of Robertson's failure lies not in these idiosyncratic events but in the general structural problems identified in the rest of this text.

Although fundamentalists are a significant minority of Americans, powerful enough to place some of their concerns on the agenda of public debate, they are not themselves united on socio-moral issues, on policies which might advance their socio-moral positions, on the value of the accommodations necessary to work in alliance with other groups, or on the priority that such concerns should have in their political choices. Furthermore, there are powerful groups which oppose them. An ironic consequence of their limited success has been to mobilize their opponents. Although this point is at present speculative, there may be a very good case for arguing that the changes in the composition and decisions of the Supreme Court, so strongly desired by the NCR, will prove to have been a Pyrrhic victory. In the last chapter of the book, I draw attention to the success the NCR has had when it has been able to present itself as the victim of élite discrimination. In so far as Robertson had a popular issue it was the claim that the judiciary had pre-empted the democratic rights of Americans by imposing unpopular liberal 'secular humanist' standards on the silent majority. The initial response of liberals was to regard any weakening of judicial support for, say, the right to abortion as a victory for the NCR. In the first place, of course, it is no such thing. It is simply a returning of a contentious issue to the political arena where contending parties will have to engage in the normal political processes of opinion formation, electioneering, and legislative haggling. It is still early days for the abortion issue but already a number of state elections in November 1989 have suggested that being in favour of abortion in certain circumstances is not the electoral liability that many feared. As I argue in the discussion of the fate of 'creation science' (pp. 118–23), increased public debate about an issue dear to the NCR may well lead, not to change in the direction they want, but to a recollection of the reasons for the status quo and a renewal of commitment to it.

Finally in this additional preface, I want to offer a partial explanation of why so many commentators have been so wrong in their assessment of the NCR. A large part of the fault is an exaggeration of the influence of the mass media. Analysts have been too ready to take the total audience for television evangelism (which is extremely hard to measure and invariably grossly inflated) and suppose that these people are NCR supporters or extremely good candidates for mobilization. Either detailed study of, or quiet reflection on, what it means to watch a television programme should have made it clear that even those who make donations to televangelism do not form a homogeneous population equally committed to all the positions that even their favourite televangelists might promote. When so many people watch so much television it is tempting to suppose that the medium has immense potential for changing opinions and hence that a socio-moral movement which has considerable access to television (as the NCR has) must itself have great potential. Even if the televangelism audience is not yet an active socio-political movement, the NCR's 'consciousness-raising' will make it so. This line of reasoning is tempting but wrong. The consensus among students of the mass media is that they reinforce existing opinion rather than radically change it and that they are better at reinforcing very general attitudes than they are at stimulating active support for particular policy positions. Access to television has undoubtedly helped the NCR (see pp. 62–4), but it has not given it and does not give it sufficient power to overcome those features of a modern democratic society which frustrate the desire to remake America as one nation under God.

S.B.

PREFACE

THIS book is a sociological account of the rise and fall of the new Christian right in America. It has a number of purposes. Taken on its own, it is an attempt to explain what is in its own right a fascinating political movement: the mobilization of conservative Protestants which can be dated from the late 1970s and which peaked with the Reagan Administration. There is always more than one way in which any phenomenon can be analysed. This is a study in the sociology of social movements. However, as with my previous writings on Protestant politics, I am more interested in demonstrating the value of sociology by its use than I am in writing about sociology. Readers who have no interest in the sociology of social movements may begin at the second chapter. I only hope that anyone who does take that route becomes sufficiently impressed with the value of a sociological perspective to turn back and read the first chapter.

This book is also the third in a series of detailed empirical studies of Protestant politics in various settings; the first two dealt with anti-Catholicism in modern Scotland and with Paisleyism in Ulster. Like them, this study has been written as a part of a project in comparative analysis, and provides material for exploring the relationships between a particular religious ideology and the social and political circumstances within which believers have to formulate their political attitudes and actions.

The material for the study comes from a variety of published and unpublished documentary sources. Although the text contains little direct quotation from interviews and little detailed description of the social world of American conservative Protestants, the most important sources were the many conversations I have had with the ordinary conservative Protestants (and their liberal opponents) and the impressions I took away from staying with fundamentalist families, attending their services and meetings, and participating, albeit briefly, in their worlds.

S.B

ACKNOWLEDGEMENTS

A NUMBER of debts were incurred in the course of the research for this book. The Nuffield Foundation generously granted me funds which allowed me to spend six weeks in America in 1983 and a further two months in 1986. On this second trip I was also assisted by an invitation to be a visiting scholar in residence at the University of Virginia. Professor Donald Black, chairman of the Department of Sociology, provided me with office facilities, and he and his colleagues made my stay pleasant and stimulating. I would also like to acknowledge the invaluable assistance of Professor Peter Metcalf of the Department of Anthropology at the University of Virginia. His spare room, veranda, company at the Bonanza Steakhouse all-you-can-eat salad bar, and collection of gruesome deities from Borneo proved excellent aides to contemplation.

The Queen's University of Belfast has been more than generous in its support of my research. While the Nuffield Foundation provided the bulk of the funds, Queen's gave me study leave to pursue my research in 1983 and again in 1986, and in 1987 it funded a third research trip to Washington and Virginia. Professor Roy Wallis and Dr Steven Yearley of the Queen's University have been patient and perceptive critics of my work and I have also benefited from the encouragement of Dr Bryan Wilson of All Souls College, Oxford, and Professor David Martin of the London School of Economics and Southern Methodist University, Dallas.

Finally, as always, my greatest debt is to the people whose actions are the focus of this study. They did not have to welcome me to their homes and offices or spend time explaining themselves to me. I am grateful that they did.

CONTENTS

ABBREVIATIONS

ACTV	American Coalition for Traditional Values
BJU	Bob Jones University
CBN	Christian Broadcasting Network
ERA	equal rights for women amendment
GOP	Grand Old Party
IRS	Internal Revenue Service
JBS	John Birch Society
NCAC	National Christian Action Coalition
NCPAC	National Conservative Political Action Committee
NCR	new Christian right
PACs	political action committees
PAW	People for the American Way
PLC	politics of life-style concern
PTL	*Praise the Lord* or *People that Love*: the Bakkers' gospel show and television network
SBC	Southern Baptist Convention
WCTU	Women's Christian Temperance Union

1

The Origins of Social Movements

THIS study of the new Christian right in America has a number of related purposes. The first and most obvious is to describe and explain what is, in its own right, an important socio-political phenomenon: the increase in political activism of conservative evangelicals and fundamentalists. The second purpose is to add to previous studies of Protestant politics in Ulster and Scotland in order to build a solid foundation for developing general propositions about the relationship between reformed Christianity and political action. Although the points of contrast will not be stressed in this study one might almost describe the political environments of Ulster and America as having opposing effects on Protestant politics. The circumstances of Protestants in Ulster give powerful incentives for even quite 'secular' Protestants to stress religious symbolism in their contest with Irish nationalists (Bruce 1986a). One of the main themes of this study is the 'secularizing' effects of the American setting; conservative Protestants, mobilized to political action on the basis of their religious beliefs and values, find that becoming involved in politics requires them to play down the religious origins of their crusade.

In both analysing the new Christian right and in laying the foundations for comparisons with Protestant political movements elsewhere, the sociological literature on social movements offers considerable analytical purchase, and at points I am interested in adding to social movement theory.

For the purpose of starting this discussion, the definition of a social movement offered by Turner and Killian will suffice: 'A social movement is a collectivity acting with some continuity to promote or resist a change in the society or group of which it is a part' (Turner and Killian 1965: 308).

Ideologies require embodiment. Ideas do not fall from heaven like the Holy Spirit of Pentecost. They are developed, preserved, and transmitted by social groups. The units deployed by macro-sociologists—classes and status groups—are too large to advance our understanding of Protestant politics much. And even less helpful is the countervailing tendency to view individuals as isolated actors. The assumption of this study is that certain approaches in the sociology of social movements offer more fruitful levels of analysis, and it is an assumption shared by many analysts of the new Christian right (NCR). Often particular disagreements about how to understand or explain certain facets of the NCR can be traced to more general differences of perspective. It thus seems sensible to begin this study of the NCR with a critical review of a number of sociological accounts of the rise of social movements, and to establish the grounds for selecting some elements of these as being useful for the understanding of the recent political interventions of conservative Protestants in America.

STATUS DEFENCE AND STATUS INCONSISTENCY

Much of the modern debate over the explanation of the origins of social movements begins with the theory of status defence, and the main analytical issues can be explored through a consideration of that approach, its antecedents, and some related alternatives. In his study of late nineteenth-and early twentieth-century American temperance crusades, Gusfield (1963) argued that moral reform campaigns can be seen as one of the ways in which a cultural group engages in political conflicts over status. If it can persuade the state to endorse its values, a cultural group simultaneously acquires public affirmation of its own status. By the last quarter of the nineteenth century, temperance had become an important point of difference between the Anglo-Saxon Protestants of the small rural towns and the later Catholic immigrants of the big cities, between the old order and the rising class. The evangelical Protestants' long-running campaign for the prohibition of alcohol enhanced the symbolic properties of alcohol and its consumption, and thus made clear the social differences between the disparagers and the imbibers of alcohol. For Gusfield, it was because the success of the prohibition campaign involved the state in acknowledging the superiority of the culture of small town and rural white Anglo-Saxon Protestants that such people supported temperance crusades.

As Wallis (1979: 96) notes, Gusfield's account initially claims to be concerned with the *consequences* of actions and not with their cause. It is a story about the 'functions' of temperance crusades, and, in places, Gusfield explicitly denies that he is presenting the *consequence* of status enhancement as a *cause* of the moral crusade. The account 'is not an assertion that this was its only function nor is it an assertion about motives. It is merely pointing out that as a consequence of such activities abstinence became symbolic of a status level' (Gusfield 1963: 59). In fact, as Wallis amply documents with quotations from *Symbolic Crusade*, Gusfield does repeat the mistake of so many functionalists. He follows an unobjectionable account of how certain consequences might have flowed from the temperance agitation with the highly contentious suggestion that it was the desire for those consequences which explained the actions. As Talmon has said of functionalist writings about millenarian religion and social change: 'Now and then there is an almost imperceptible shift from imputation of latent functions by the investigator to imputation of self-conscious distortion and disguise by the actors' (1969: 253).

In a critique of such reasoning, Wallis establishes firstly that even uncontentious descriptions of what consequences follow from an action cannot, without further independent evidence, be used as explanations of that action. Hardly surprisingly, unintended consequences that are unpleasant are rarely turned by sleight of hand into causes. Tripping over and breaking one's nose is not often offered as a cause of running down unlighted stairs. But even pleasant unintended consequences cannot be so used. It is not yet an explanation to show that a group, the status of which was being threatened, engaged in some action which, had it succeeded, would have had the latent function of enhancing the group's status. Given that motives other than a desire for status enhancement (such as altruistic concern for the moral and economic welfare of drinkers) remain a possibility, one would have to produce independent evidence that enhanced status was what was actually desired, and was not simply an unintended consequence. Is it the case that the diaries, letters, speeches, and publications of temperance crusaders display a concern with their social status? Gusfield presents no such evidence. In its absence, Wallis (1979: 98) argues, we can conclude three things:

1. [the temperance campaigners] are dissembling and *really do* desire status improvement and know this to be the case, although they will not admit it.

2. They are motivated by unconscious desires for status improvement which they fail to recognise.
3. The hypothesis of status discontent is false.

Gusfield presents no evidence for (1) or (2), which strongly disposes us to conclude (3).

But there is an even greater problem which is common to many of the cases where this sort of reasoning is used to explain social movements. Not only is the underlying logic questionable, but the descriptive data also fail to establish a necessary plank of the argument. Gusfield does not even present compelling evidence for the claim that his temperance crusaders shared a common status in the first place. His own data show that many Women's Christian Temperance Union (WCTU) activists were actually drawn from the labouring classes. Given the common tendency for movement activists to be of higher social status and occupational class than ordinary supporters, this suggests that many WCTU members were not members of the threatened old middle class.

Despite these problems, Gusfield's theory of status defence became popular for explaining movements with non-economic goals and it is now deployed to explain the new Christian right. In the 1970s it was adopted by Zurcher and his associates to explain two anti-pornography crusades. Their use of status defence is even less plausible than Gusfield's because they have greater trouble demonstrating that the actors in question shared a common social status (Wallis 1979: 102). Their solution is to complement the status defence approach with a theory of *status inconsistency*.

> The status inconsistency perspective can be traced to Lenski's (1954) proposal that individuals who have discrepant or inconsistent ranks on status dimensions, e.g. high educational achievement and low occupational attainment, may be subject to discontent and stress that is reduced by compensatory attitudes or mitigating behaviors. Such attitudes and behaviors represent the adoption or intensification of a response pattern that is considerably beyond a 'middle of the road' position. (Simpson 1985: 156)

Finding themselves in a psychologically unstable position, status inconsistents will express their status anxieties by engaging in irrational episodes of collective behaviour. They will be in the market for various emotional and irrational appeals. Rather than following the institutional channels for the expression and redress of grievances—the normal ways of doing political business— they will be recruited to enthusiastic social movements. Those who join anti-pornography crusades are thus not to be seen as rational people acting in a way which they believe will lead to the banning of pornography. Instead they are irrational actors expressing their status anxieties. That is, to return to the term which Gusfield popularized in the title of his study of the WCTU, their politics will be 'symbolic' rather than realistic.

In their time, status inconsistency theories have been used to explain right-wing extremism (Eitzen 1970a), liberalism (Hunt and Cushing 1970), drinking behaviour (Parker 1979), aggression (Galtung 1964), suicide (Gibbs and Martin 1958), susceptibility to illness (Jackson 1962), and seeing flying saucers (Warren 1970). Little of the status inconsistency research is convincing. Numerous conceptual problems have been identified and subsequent attempts to replicate findings have failed (Box and Ford 1969; Wallis 1979; Olson and Tully 1972; Jackson and Curtis 1972). Unable to find any strong correlations between gross inconsistencies on education, income and occupational prestige, and involvement in an anti-pornography campaign, Zurcher *et al.* are forced to rest their argument on a weak second-order correlation: 'Individuals who are highly rewarded with income but who are education–occupation inconsistent, are more likely to participate in change-resisting social movements than are education–occupation consistents' (Wilson and Zurcher 1976: 530).[1] In many other studies which have tried to use status inconsistencies, the amount of variation in whatever is being explained which can be correlated with status inconsistency is usually less than that which can be explained by any of the statuses taken by themselves (Eitzen 1970b).

But what is most disturbing about this research is its inability to construct a convincing explanation of why status inconsistents should be attracted to anti-pornography campaigns. In a detailed exploration of the nature of sociological explanation, Wallis and I have argued that relationships between 'structural properties', even

[1] It is worth adding that the data collected in this and similar surveys are extremely shallow. Even if income, status and education are important, the operational definitions of these things are superficial. What people say they earn is only part of income. How many years of formal education someone has had is at best a crude measure of education. It cannot be repeated often enough that no amount of sophisticated statistical manipulation improves shallow data.

relationships of constant concomitance could such things be found, are not yet explanations.

Our view is that we have no ways to understand what such correlations mean, or how they operate causally, except through an action story. When such an action story is secured, the explanation lies there, not in the structural correlation. (Wallis and Bruce 1986: 35)

Even if we accept that Zurcher *et al.*'s claimed second-order correlations between status inconsistency and moral crusading are not simply artefacts of the measurement process (and, given the contrast between the crudity of the original data and the sophistication of the statistical analyses, they might be just that), we still have the problem of imagining what interpretative procedures have led actors to be influenced to join anti-pornography campaigns by status inconsistencies so elusive that trained sociologists have trouble finding them.

Lenski's original formulation at least had some intuitive appeal. He supposed a 'game theory' view of social interaction in which each player tries to maximize his own advantage while minimizing the advantages of the people with whom he interacts. Status inconsistents will see themselves in terms of their higher status while others will see them in terms appropriate to their lower status. Hence they will have an uncomfortable life.

As Box and Ford (1969) argue, there are a number of tenuous assumptions even in this formulation. It supposes that people *know* about each other's status vulnerabilities and that people do not 'pass' as being higher than they 'objectively' are on their weak dimension.[2] It also assumes that people actually, frequently, and seriously play the game of status contests. But the greatest weakness is that, for Lenski's formulation to work, it would have to be the case that ordinary people not only shared sociologists' interest in status but also operated with the same model of status. Attempts to discover among ordinary people a subjective sense of status inconsistency have failed miserably. The conclusion of one such study was that:

The extremely low correlations between the perceived inconsistency measures and other measures, objective or subjective, lead us to doubt that respondents, left to their own [*sic*], would employ modes of status comparisons which would be the subjective analogue of [sociologists']

[2] For a good illustration of ways of 'passing' see Goffman (1970: 92–112).

operational procedures for standard measures of status inconsistency. (Starnes and Singleton 1977: 263)

Furthermore, one of the few detailed attempts to discover how ordinary people view occupational prestige produced models of stratification and status interestingly different from those assumed by sociologists (Coxon, Davies, and Jones 1986).

If the Lenski formulation has at least the semblance of sense, that is more than can be said for the Zurcher *et al.* attempt to produce a convincing story to link their second-order correlations and involvement in anti-pornography crusades.[3] They tell us that people who are financially 'over-rewarded' (that is, those who are paid more than their educational attainment or occupational prestige would seem to merit) will resist changes in their social environment which threaten their precarious over-rewarded position. But it is not clear why being 'over-paid' should be precarious. It is not even clear that 'being over-paid' is a meaningful notion in a culture as success oriented as that of America. It is even less obvious why, even if such people did feel that their luck was a little too good to be true, they should feel threatened by pornography. As with Gusfield's study, the Zurcher *et al.* work produces no evidence that the actors in question *felt* their status to be under threat (Wallis 1979: 101).

What makes the failure of status inconsistency and status defence explanations to produce convincing evidence all the more noticeable is their neglect of the ample evidence for an alternative 'threat' story which is present in the stories which temperance and anti-pornography campaigners themselves tell. Although they are determined to ignore its implications, Gusfield and Zurcher *et al.* actually show that the objects of their studies did feel threatened. However, it was their *culture* rather than their social status which the crusaders felt to be in danger.

THE NEGLECT OF ACCOUNTS

The neglect of actors' accounts is not confined to status defence and status inconsistency approaches to social movements. It is such a common feature of what was, until the 1970s, the dominant tradition in the social movements literature that even those theories

[3] In a general survey of the literature on social movements, Zurcher (Zurcher and Snow 1981) later admitted the weakness of status inconsistency theories.

which claimed an interest in actors' perceptions managed to ignore what actors said. Commentators who appreciated that the most deprived and most persistently oppressed sections of a population were often the least likely to rise in rebellion promoted the importance of 'relative deprivation' as an alternative to absolute deprivation as a source of collective action. However, although such work often entailed a programmatic commitment to actors' perceptions, it frequently made no effort to discover how people felt about their circumstances. It has been rightly said of Gurr, for example, that 'he does not often cite relevant data about people's perceptions when he discusses relative deprivation, even though perceptual data are implied by his general definition, and appear in his operational definitions. Instead he frequently uses "objective" data, such as employment rates, to infer the psychological state of relative deprivation' (Marx and Wood 1975: 378).

There seem to be two basic reasons why the possibility which accords best with actors' accounts is discounted in favour of an explanatory route which stresses irrational response to objective pressures: meta-theoretical disposition and methodological ease. As Wallis argues in the case of Gusfield, many analysts of social movements are 'closet' functionalists. If the cause of collective behaviour is to be located in some latent function, it is not to be expected that the actors would be aware of that cause, and hence we should not expect to find evidence of it in actors' own appreciations of their motives and intentions. Even when the explanations offered are structuralist rather than functionalist (that is, when the action is a response to the stimulus of some structural stress, whether or not that response may have some suitable latent functions), such explanations still assume that people are caused to act as they do by social forces outside their knowledge. For such approaches, the only truly sociological explanation is one which reduces actors to being simply or largely the carriers of structural properties.

There is also a methodological reason for neglecting actors' accounts. It is easier to collect 'objective' data than it is to talk to people at sufficient length to find out what they are doing and why. Structural properties, such as years of education, occupational status, and income can be quantified and presented as, to use a common American expression, a 'data set'.[4]

[4] An additional reason for the neglect of actors' accounts is that reliance on them may give sociological analysis the appearance of being little more than common

STRUCTURAL STRAIN AND IRRATIONALITY

Underlying both meta-theoretical dispositions and methodological considerations is the unfortunate legacy of the earlier structuralist writings on social movements of Parsons, Smelser, and Lipset. Parsons said: 'It is a generalization well established in social science that neither individuals nor societies can undergo major structural changes without the likelihood of producing a considerable element of "irrational" behavior' (1969: 169).

Leaving aside the questionable confidence that any generalization in social science is 'well established', we should note that, thereafter in this work, the word 'irrational' appears without the distancing effect of the inverted commas. Structural strains cause 'conspicuous distortions of the patterns of value, and of the normal beliefs about the facts of the situation' (1969: 169). The people affected suffer high levels of anxiety and will direct considerable aggression towards the supposed 'source of strain' (1969: 170).

The same basic dichotomizing of rational and irrational action is found in Smelser's (1966) model of social movement genesis, which suggests at least three initial requisites:

1. The existence of a social strain of ambiguous proportions which creates widespread anxiety.
2. The designation of a specific cause for that strain. Without such a designation, there would be no movement, only hysteria.
3. The designation of a specific solution. Without such a specific designation there would be no movement, only a kind of wish-fulfillment. (Lipset and Raab 1978: 23)

The original impetus to collective action is the causal connection between structural strain and anxiety. Although he is less free with the word 'irrational' than is Parsons, Smelser clearly shares the view that social movements are to be seen as distinct from, for example, conventional political party activity, not only in being 'uninstitutionalized', but also in being caused by anxiety rather than being motivated by reasonable responses to the situation. As Parsons put it, there are 'normal beliefs about the facts of the situation' and

sense. As Blumer (1967) pointed out, many social scientists place more weight on imitating the methods of the most advanced physical sciences than they do on adopting methods suitable to the nature of the empirical world they study. Like other occupational groups seeking professional status, sociologists have felt obliged to distance themselves from the methods of the laity. The mistake was to do the distancing on what counts as 'data' rather than on the rigour with which it is evaluated.

there are 'conspicuous distortions' of such beliefs. Social movements are produced by the latter rather than the former.

Unfortunately (given the considerable merits of other parts of his political sociology), Lipset follows Smelser and Parsons. Lipset is concerned with political extremism, rather than with social movements *per se*, but there is considerable overlap of the two fields and his work has been sufficiently influential to deserve discussion. For Lipset, political extremism is defined by its *monism*: its unwillingness to accept the democratic principles that people have a right to be wrong and that the tolerance of conflicting ideas is a virtue. The extremist is likely to 'favour a simplified view of politics, to fail to understand the rationale underlying tolerance of those with whom he disagrees, and to find difficulty in grasping or tolerating a gradualist image of political change' (1983: 108). In his introduction to detailed studies of various right-wing movements in America, he endorses both Smelser's model and Hofstadter's view that such movements are characterized by a 'paranoid style' (Lipset and Raab 1978: 12–20). The reliance of many movements on a conspiracy theory of history—the idea, for example, that American foriegn policy failures in the Cold War period were caused by a secret communist conspiracy in the Army and Pentagon—is taken as evidence that involvement in such movements is to be explained as an 'over-determined' response to the stimulus of structural strain, rather than as a relatively reasonable attempt to understand the situation. In a related essay on working-class authoritarianism written in the early 1960s (1983: 87–126), Lipset invokes many of the ideas of the 'authoritarian personality' approach to extremism. The working classes' lack of education, relative isolation from cosmopolitan culture, child-rearing practices, and economic and emotional insecurity are all offered as roots of a general susceptibility to simplistic stories about the causes of problems and their possible solutions. A decade later, the personality approach was down graded and greater emphasis was placed on the various structural conditions which create different sorts of anxiety. Extremist movements of the lower classes tend to be driven by economic anxieties. They wish for the power and status that have been denied them, which they believe to be their right, and which they wish to gain through the actions of the state. Extremist movements of the lower middle classes tend to be driven by a desire to *regain* the power and status which they have recently lost or which they fear they will lose. Such movements tend to be anti-statist. More sophisticated analyses of structural changes are introduced to specify further the nature of the sorts of extremist movement which such changes will produce.

Lipset's analyses cover so much substantive material that his work cannot and should not be easily dismissed. Furthermore, as my interest is only in those social movements which pursue largely non-economic goals, there is no need to engage in comprehensive criticism of his work. However, his approach to 'symbolic politics' is so obviously relevant to Protestant political movements—and he has himself written at length on such movements—that elements of it must be considered.

There are a number of objections to a status anxiety explanation (although there may be particular cases where it can be demonstrated to be valid). One problem is inadvertently demonstrated in Vander Zanden's explanation of recruitment to the 1950's Ku-Klux-Klan. Klan members in the area studied are divided by occupation into four groups: skilled workers; small business men; 'marginal' white-collar workers; and semi-skilled workers in industries such as transport and construction. Vander Zanden happily describes the first three groups as insecure and status anxious (1960: 458). While one can undoubtedly construct a sociology to justify such a description, one runs into this problem: in such a model, what social groups are not status anxious? If the deprived are anxious about their lack of status, and the majority of those with status fear losing it, who is left contented and secure? Presumably the only groups which are not likely to provide recruits to extremist movements are those that have quite recently seen a considerable improvement in their status and that have not yet started to worry about losing it.

Anyway, there are even more basic problems in maintaining the image of members of extremist movements as low status, ill-educated marginals. Stone's study of the John Birch Society shows the Californian Birchers to be: 'for the most part, well-educated, reasonably young individuals with substantial family incomes. The men are employed in upper-status occupations, although they are rarely self-employed' (1974: 185). Similarly, supporters of Fred Schwarz's Christian Anti-Communism Crusade in the early 1960s tended to come from 'an upper status group. More than half are business or professional people, 41 percent with incomes of more than $10,000 a year. More than half are college graduates. . . . These citizens are not crackpots or malcontents, nor are they lonely, frustrated individuals looking to join just about anything that meets and screams' (Forster and Epstein 1964: 58–9). All of which brings us back to the problems of Gusfield's study. If one cannot demonstrate either (a) that those recruited to symbolic politics share a common status or (b) that the status they share is one which is especially likely to be accompanied by a sense of anxiety, then one has to move to more elaborate formulations such as status inconsistency, the weakness of which has already been discussed.

The most abstract objection to Lipset's theory of status anxiety and other structural theories of social movements is that they begin by denying normal motives to people who engage in uninstitutionalized activities. People who engage in 'extremist' movements are seen as being determined to act in an expressive and irrational fashion by the anxiety which results from structural strains. Such theories rule out the possible alternative that people who accept conspiracy theories of the world, for example, do so because such theories appear reasonable and plausible. In his discussion of working-class authoritarianism, Lipset notes the common association of millenarian religion and political extremism, and sees both as evidence of what are essentially personality disorders. An alternative way of seeing the connection is to suppose: (a) that traditional supernaturalist religions are themselves 'conspiracy' theories with God and the Devil—good and evil—as the main conspirators; (b) that people who are raised in such religions accept their basic cognitive styles; and (c) that, for such people, conspiracy theories are reasonable ways of interpreting the world. To return to Parsons, the interpretative sociologist doubts the explanatory value of asserting that any one particular view of the 'facts of the situation' has such obvious validity that alternative views can be dismissed as the product of unusual, hysterical, and irrational interpretative procedures. To say this is not to endorse extremist world-views. It is simply to say that the explanation of why people believe something cannot be bound to the truth or falsity of the belief in question. There may actually be a God. There may actually be some divine providence which makes sense of the apparent anarchy which surrounds us. There is nothing which the social scientist knows which gives him or her an insight into such questions any greater than that possessed by the average Klansman or John Bircher. Hence no system of social scientific explanation can be based on a distinction between true and false belief.

Parsons talks of the true facts of the situation. Smelser, Lipset, and Hofstadter could reply that they do not need to be able to determine truth and falsity to establish the rationality and appropriateness of any particular set of beliefs. Thus they could accept that some of the things which liberal middle-class people presently accept to be true will turn out to be false but still maintain that those beliefs are 'rational' in that they accord with the best evidence presently available. The beliefs of the average Klansman do not so accord. Therefore they are irrational. As such, they can be explained in a manner different from that used to explain the beliefs of a Lipset or a Smelser. Wallis and I have argued that this reworking does not resolve the problem. One does not have to be a committed relativist to recognize a large element of social construction and convention in what will count in any particular culture as 'good grounds' for any particular belief. While we do not discount the possibility of sensibly describing some belief and action as irrational, we argue that most cases of apparently curious behaviour will turn out to be 'reasoned' once one learns a lot about the people in question. To return to the example already suggested, a belief in conspiracy theories may seem strange to an atheist but it seems quite sensible to some people who already believe that there is a hidden order in the world, that history has a purpose, and that Satan is a reality.

A related problem results from conflating the institutional/uninstitutional and rational/irrational dimensions for the description of social action. There seems little warrant for treating the two dimensions as if they were always, or even usually, linked. Is there really all that much difference between the National Front's view that Britain's economic, social, and political decline is all the fault of the presence of non-white immigrants and the Thatcher Government's view that monetarist control of the money-supply as defined by the Sterling M3 measure will reverse the economic decline? Is there a clear difference of kind between the fascist belief that all Britain's problems are a result of a communist plot and the Thatcher Government's often stated (and acted on) belief that inflation and poor productivity are caused by 'the power of the unions'?

It is often not clear whether a certain course of action is judged to be irrational because it is based on distorted beliefs about the real world or because it was never likely to achieve its ends. The problems with the first characterization have already been mentioned. The problem with the second, of course, is that it is only *after* the event that we know if a certain plan of action succeeds. Control of Sterling M3 did not provide a useful device for controlling the economy and it was eventually dropped. But that it did not work is not the same as saying that any reasonable person should have known it would not work, and hence that those who believed that it would were irrational, either in their beliefs or in their acting upon those beliefs.

In many particular cases, the notion of status anxiety is not only pejorative but also unnecessary. The data on what sorts of people joined certain movements can be used to support a more straightforward and 'realistic' explanation of their actions. The Ku-Klux-Klan, for example, 'recruited from among those who reacted negatively to the social strains of rapid growth, changes in class relations, etc. (another possibility is that it drew its support from recent migrants from smaller communities)' (Lipset 1965: 78). Detailed studies support the claim that the Klan had considerable appeal to Protestants who had recently moved from rural villages and small towns to the growing industrializing cities of the piedmont areas of the South and other parts of the South West (Moffat 1963; Rice 1972; Alexander 1965). One way of conceptualizing the actions of those who joined the Klan is to suppose that the social strains involved in such a move (and in the restructuring of social relations for even those who did not move) created anxiety which people both expressed and sought to allay through involvement in extremist movements. But it seems clear that the story works every bit as well if the 'strain–anxiety' axis is omitted entirely. What one is left with is a population of people who did not like the way the world was changing and who acted to demonstrate their opposition to those changes.

It often seems that the depiction of social movement activity as irrational rests, not on any warranted analytical distinctions, but on taking for granted the dominant beliefs of middle-of-the-road cosmopolitan intellectuals. Like much sociology of religion, much sociology of social movements is a 'sociology of error' (Hamnett 1973). Smelser's distinction between instrumental and expressive action is reflected in status defence and status inconsistency explanations of symbolic politics. 'Symbolic' is not so much a term of analysis as a sign of the assumption that no one in their right mind would actually believe it important to campaign for temperance or censorship of pornography, and hence that those people who do engage in uninstitutionalized action in pursuit of such goals must be driven by something other than conventional sources of motivation. It is not good reason but anxiety caused by structural strain, status inconsistency, or some such which explains collective action.

Such reasoning seems indefensible. Just as we cannot follow functionalists who read back from 'successful' consequences to cite the desire for such consequences as the cause of the action, so we cannot take the failure of any particular campaign as evidence that the actors themselves were not behaving rationally and instrumentally. Hence, we cannot divide action into two discrete types—rational/instrumental and irrational/expressive—and use diferent methods to explain these types of action. Without good evidence to do otherwise (and the suspicion that all people should know that collective action does not work is not good evidence), we must practise 'explanatory monism' and treat all social action in the same manner. Elsewhere Wallis and I have argued in detail that the explanation of social action rests on the discovery of intentions and motives. As sociologists, we are committed to the belief that intentions and motives are social, rather than idiosyncratic, productions and that they will, therefore, often be related to shared 'objective' characteristics such as class and status positions. But, while they are related, they are not reducible to such characteristics. People do not carry and exhibit symptoms of social characteristics in the same way that they carry a virus. All social action is mediated by interpretative processes.[5]

CULTURAL DEFENCE

Fortunately, not all American analysts have pursued the irrationalist approach to the genesis of social movements. A number have chosen to take seriously actors' accounts and have revived a perspective closer to Weber's writings on status than the status

[5] Most sociologists pay lip-service to this notion but entirely neglect it in their practice. Sadly, little has changed since Blumer's excellent critique of the neglect of interpretative procedures.

defence stories for which he is claimed as an intellectual progenitor. They have observed that it is not personal social status which moral crusaders and their like were moved to defend. Rather, as Wallis (1979: 102) put it, it is the status or prestige of a particular culture which is at issue. To describe their actions anachronistically, Gusfield's temperance crusaders were not defending their own social status but the status of a life-style. In the American social science literature, this cultural defence approach acquired, through Page and Clelland's study of controversy in the American rural South over textbook content (1978), and Lorentzen's analysis of opposition to the equal rights for women amendment to the American Constitution (1980), the rather inelegant title of 'the politics of life-style concern' or, for the sake of brevity, PLC. In this view, a movement such as the new Christian right would be explained as a reaction to changes in the social, moral, cultural, and political environment which threatened to undermine the ability of a particular group to maintain its shared culture.

One would expect structuralists to criticize the PLC approach's rejection of structural strains and status anxieties as major sources of impetus to collective action. However, the PLC perspective has also been attacked from the 'left'. Some analysts who share the above doubts about structuralist approaches of social movements have noted that Smelser, for example, talks about grievances and motives, and have mistakenly taken such structuralist versions as representative of what will be produced if one concentrates on motives. Failing to notice that the structuralists have a mechanical and essentially irrationalist view of the link between problems, grievances, and collective action, the 'left' critics have determined to eschew entirely talk of motivation.

Often described as 'resource mobilization' theorists, scholars such as McCarthy and Zald have argued that one cannot explain the origins of social movements by pointing to some grievances which the supporters of the social movement are attempting to rectify or resolve because such grievances are omnipresent. Being more of a constant than a variable, 'grievance' cannot be the cause of a social movement. What explains the rise of a particular movement organization is the existence of some skilled cadre which capitalizes on a pre-existing sense of grievance to mobilize resources.

In a recent attempt to enhance our understanding of the new

Christian right, Miller (1985) has criticized previous PLC work. Firstly, he points out that we cannot simply assume that the statements of movement leaders about their reasons for activism can be taken as representative of the motives of rank-and-file participants. Thus when the Revd Jerry Falwell tells us that he became politically active in the late 1970s because the moral degeneration of America had gone so far as to demand an active response, we cannot assume that rank-and-file new Christian rightists shared a similar perception. Secondly, Miller notes that we cannot assume, just because NCR activists are *now* concerned about their culture, that they were not equally concerned at various times pre-dating the rise of the new Christian right. Hence he argues quite sensibly that, if we are to explain the genesis of the NCR as a response to an increased threat to the culture valued by conservative Protestants, we need longitudinal data about the interests and concerns of the sort of people who support the new Christian right which would show an increase in the sense of cultural threat.

The first point is a useful caution but it is not a devastating criticism of earlier work. We certainly cannot assume that movement members all share a common perception which mirrors that of the leadership, but nor need we assume infinite variation between the interests of leaders and members. Activists who signally fail to advocate views which resonate with the perceptions of potential members will fail to build and sustain followings.

Miller's criticism of Lorentzen for not presenting longitudinal data is also valid but not crushing. We know that accounts presently given by actors are not only attempts to retrieve the past accurately. They are also influenced by the circumstances in which the account is given and by considerations of the future. For example, we have good reason to suppose that recently 'born again' Christians exaggerate the sinfulness of their pre-conversion lives. Their view of the past is heavily coloured by their new world-view and by expectations built into the role performances customarily associated with giving one's testimony. But almost all social investigation relies to some extent on people's recollections of the past and although we should always be sensitive to 'rewriting', there is no meta-methodological reason why, for example, we should not believe Schneider's North Carolina fundamentalist pastors when they told him that they became involved in politics in

1980 because only then did they realize how far the country had gone to the dogs (Schneider 1986). It might, of course, be the case that one's respondents are lying, ignorant, or themselves deceived but these are problems for each particular event of account giving and need not form the basis for a wholesale rejection of people's memories. Furthermore, it is incumbent upon those analysts who deny the validity of images of the past presented in currently given accounts to offer good evidence for their view. It would, of course, have made Schneider's job easier if he had had on record stories from his pastors' pre-politicization phase but there is no reason to suppose that actors' accounts are invariably tied to the occasion on which they are given (Bruce and Wallis 1983; Wallis and Bruce 1986: ch. 1).

But these are points which can only be settled in particular cases by the evidence, and the movement we are here concerned with is the new Christian right. Miller analyses the contents of sample issues of long runs of six leading evangelical publications and concludes that concern about life-styles and evangelical culture is nothing new. Instead of there being any marked increase in the number of articles discussing threats to fundamentalist ways of life or in the intensity of calls to political action, Miller finds that jeremiads have been a regular feature of conservative Protestant literature, as have calls to some sort of oppositional political activity. He concludes that the PLC approach to the new Christian right, having been thus tested, is found wanting, and offers as an alternative the resource mobilization idea that the NCR is explained, not by a new sense of concern among fundamentalists, but by new methods of sensitizing and mobilizing the saints. I have offered detailed criticisms of Miller's method of 'content analysis' elsewhere (Bruce 1987). Here I will confine myself to a few of the more general and obvious weaknesses in this test of the PLC explanation.

A general weakness is that Miller has nothing to tell us about how the readers receive the publications in question. To give a trivial but none the less illuminating example, I have taken the same daily paper for fifteen years. Only in the last five years have I started to read the financial pages and to do the crossword. Both those features have always been there but my attention has shifted. Miller's research tells us nothing about how readers attend to the various parts of the publications he analyses. Even more disturbing, he does not tell us what has happened to the circulation figures for the serials under discussion. Do they have more or fewer readers now than in 1972? Carl McIntire, the founder of the fundamentalist American Council of Christian Churches and International Council of Christian Churches (conservative alternatives to the comparable ecumenical bodies), has been flogging the same anti-Catholic and anti-communist horses for thirty years. What is important is the direction and extent of the change in the number of people who accept and act on his analysis of the world and these are not things which can be discovered from content analysis.

But even if we allow that the incidence of certain topics in Miller's six serials is a regular reflection of the interests of the readers, the analysis of such incidence would only test the PLC thesis: (a) if the readership was representative of NCR supporters; and (b) if the interests and concerns of the readers were *fully* and consistently displayed in their reading of these serials. These conditions are not met by asserting that the six serials were chosen randomly; that is not the problem. The difficulty is this: what if a regular reader of the *Moody Monthly* began, for example, in 1980 also to subscribe to more overtly political publications such as Falwell's *Fundamentalist* or the *Moral Majority Report*? That would be a good indication of a conservative Protestant becoming more concerned about threats to his or her culture. Were it repeated on any scale, it would be precisely the sort of evidence that would support a PLC approach. Yet it is a possibility which Miller's method cannot explore.

If one were genuinely interested in using periodicals as guides to the interests of conservative Protestants, one could approach the problem from one of two directions. One could consider the market as a whole and compare the circulation fortunes of a large variety of more and less overtly politicized periodicals. Or one could start with the readers—the people whose behaviour is supposedly being explained—and interview a large number of them, at length, about what magazines they read, what sorts of articles catch their attention, and what issues cause them concern.

But even this would only be a start. Why suppose that the interests of conservative Protestants can be gauged from the content of what is only a part of their reading matter, and when reading is only one of the ways in which they learn about the world and how they should react to it? Miller's method assumes that evidence of the characteristic he is seeking— increased concern about socio-moral issues—should be regularly distributed through all channels of communication, and thus should appear in these serials. Given that increased concern about socio-moral issues could equally well be evidenced by subscribing to new periodicals, either as replacements or as supplements, by increased audiences for the more overtly political televangelists, or by a new militancy in the pastorate, I see no reason to accept Miller's assumption, at least not so whole-heartedly as to follow him to his conclusion that the PLC approach is deficient.

What we need to know to explain the genesis of a popular social movement (as opposed to one which relies on strong initial funding to generate the *appearance* of popular support) is why people become involved. The obvious thing to do is to ask them. Unless one has a convincing and sustainable theory to explain why people should be less knowledgeable than we are about their motives, we should solicit their accounts of their actions. Unless the actors in question are no longer available for questioning, anything else can only be justified as preliminary or secondary investigation. Clearly the best way of asking people why they acted as they did is the face-to-face interview. In addition to direct interviewing, there are sources of data which come close to 'asking people'. Since the publicity surrounding the launch of the Moral Majority in 1979, many skilled journalists have researched lengthy articles on the NCR and its supporters: articles which contain considerable interview material (see, for example, Fitzgerald 1986 or Furguson 1986). Although there are a number of weaknesses with attitude surveys, a considerable number of studies of NCR involvement based on such surveys now exist. At a number of points in my discussion I will point out lacunae in the evidence but considerable material concerning the motives and intentions of NCR activists and supporters is available, even if it is not readily compressible into quantitative 'data sets'.

SOCIAL CHANGE, GRIEVANCES AND MOBILIZATION

Resource mobilization explanations of social movements begin by being agnostic about the relationship of motives to actions in the genesis of social movement organizations. Or, at least, they are agnostic about the motives of rank-and-file members. If we assume that a sense of grievance about any issue is widespread, we need not look for an explanation of a new sense of grievance which makes people potential recruits for a movement directed at the amelioration of that grievance. Instead the genesis of any new social movement organization is to be explained by pointing to the ways in which 'resources'—members, funds, publicity, and so on— have been mobilized. Resource mobilization is clearly central to a complete explanation of social movement phenomena and it is a useful corrective to those theories which have taken the connection between 'grievance' and 'response' to be unproblematic but it cannot stand on its own. In seeking to escape from earlier models which supposed that, once one had shown the source of some grievance, one had explained why people were recruited to the appropriate social movement, resource mobilization has gone too far in the other direction. The mistake seems to stem from a misreading of the 'tradition' against which resource mobilization is defined as the alternative. McCarthy and Zald correctly say of 'traditional' social movement theory that it assumes that '[a]n increase in the extent or intensity of grievances or deprivation and the development of ideology occur prior to the emergence of social movement phenomena' but they are mistaken when they go on to say that the main tradition 'has emphasized even more heavily the importance of understanding the grievances and deprivations of participants' (1977: 1214). Far from it. As the above discussion should have made clear, the main flaw with early social movements research was precisely that, instead of trying to understand the grievances of movement participants, researchers blithely assumed that activists 'responded' to structural stimuli. And although many analysts assumed that both grievances and ideology preceded mobilization, there is no pressing reason why the two things should not be seen as interacting. After all, people are not going to support a civil rights movement unless they (or some other people with whom they sympathize) feel they have been deprived of their civil rights. Equally well a shared ideology is usually required to convince people to see their situation as being problematic, as something which can meaningfully be constructed as a 'grievance'.

A moderate statement of a resource mobilization view—'grievances and discontent may be defined, created and manipulated by issue entrepreneurs and organizations' (McCarthy and Zald 1977: 1215)—is perfectly acceptable and it is hardly novel; symbolic interactionists such as Howard Becker were saying as much in the

1960s. In the exaggerated form suggested by Miller, resource mobilization's neglect of the motives of the rank and file brings it perilously close to the more mechanistic Marxist and Machiavellian models of political action. It is also clearly not the case that grievances are a constant rather than a variable. The world does actually change. While Miller is right to suggest that conservative Protestants have always found things to complain about, he is wrong if he supposes that they have all had equal salience. Conservative Protestants of the 1950s were offended by girls smoking in public. In the late 1960s girls were to be seen on newsfilm dancing naked at open-air rock concerts. While both can be abstractly construed as culture threats, the second was taken to be a much greater threat than the first. In the 1940s and 1950s fundamentalists were exercised by the rather abstract and distant threat of 'world communism'. By the late 1970s they were faced with the added and apparently more immediate threat of an increasingly secular state which acted, through the Internal Revenue Service for example, to reduce the autonomy of fundamentalist institutions.

I have already suggested that some confusion can be cleared up if we distinguish different types of social movement organization according to the extent to which they involve popular participation. Some contemporary agitations involve almost no rank-and-file support. Instead a small cadre of professional organizers solicits funds from charitable trusts and from the government, and uses such funds to create publicity favourable to their campaign. Such PR work can be used to create the impression of popular concern and so have the intended effect of influencing legislators or some government body into taking the desired action. If one thinks of these as top-down movement organizations, there are still bottom-up organizations, the more traditional movement in which large numbers of people are recruited to support a campaign. As I will make clear shortly, the new Christian right involves elements of the top-down model, but it has also recruited large numbers of ordinary people to support it. If one is looking for evidence of popular involvement, one need only consider the way in which the NCR raises money—in small sums from large numbers of people—and exerts pressure—by registering new voters and persuading them to vote for particular candidates in elections and for propositions in referenda.

For all its motivational agnosticism, resource mobilization does suppose that the cadre of activists is motivated. The explanation of why some people become activists often rests on changes in their relevant environment. For example, there is now a range of anti-smoking organizations which uses professionals to create the appearance of representing a major social movement. These professionals became involved because they became convinced by the evidence that smoking was extremely bad for one's health. It seems curious to be willing to permit the cadre of activists to have been stimulated into action while denying the same right to rank-and-file members of popular social movements.

The popular involvement in the NCR suggests that many ordinary people did believe that they had some grievance which deserved attention. My argument does not require that fundamentalists were entirely happy with their world until certain recent changes threatened it. All it requires is that there be an identifiable increase in the intensity of such threats at some point prior to the growth of the movement. I will also add a previously neglected element. It may well be that what changed in the 1970s was not *just* the degree of threat to the culture of conservative Protestants but also their beliefs about the possibility of making effective protest against such threats. Irrationalist approaches to social movements, which see them simply as expressive responses to social change or stress, neglect the possibility that an episode of collective action may have its roots in an increasing sense of optimism about the possibility of effective action. If movements such as the NCR are bracketed as 'symbolic' politics, search for the change which caused such movements will be misleadingly confined to those changes which weakened the position of those who became active. If, however, we avoid the assumption that moral crusades are irrational responses to social stress, we can entertain the possibility that such movements occur at the point where two social currents meet; at the point where a group feels sufficiently troubled about something to want to act *and* feels that such action might work.

A resource mobilization approach is actually needed for both these trends. Shared objective circumstances do not automatically produce shared responses. People have to communicate amongst themselves in order to construct a shared definition of the situation as being problematic and as being amenable to rectification. Leaders—those people engaged in attempting to mobilize resources—play a vital role in stimulating and directing such communication.

If this is accepted, it follows that a full explanation of the rise of the new Christian right must involve three elements. It should firstly explain why there was a market for such a movement. Secondly, it should explain how potential supporters were sensitized, politicized, and mobilized. This will be a story about leadership. Thirdly, because neither leaders nor followers act in a vacuum, it will consider the social and political structure within which this mobilization took place.

2

Genesis: The Market

In order to understand why the 1970s should have seen the emergence of a potential market for something like the new Christian right, and why the movement took the form that it did, we must briefly consider certain aspects of the history of Protestantism in America and certain features of the political environment.

The United States of America was once a Protestant country: 'colonial religion generally derived from the tenets of the Protestant Reformation' (Carroll and Noble 1982: 62). Many of the settlers were self-consciously Protestant, having moved to the new country so that they could practise their dissenting religion without hindrance. Those who did not see themselves as persecuted saints were nevertheless from Protestant cultures. As late as 1830, only some 300,000 people (about 3 per cent of the population) were Roman Catholic (Carroll and Noble 1982: 179). That most early Americans were not Roman Catholic did not, however, produce an homogeneous religious culture because the diverse ethnic origins of the settlers were reproduced on American soil:

> The tendency to form ethnic enclaves, which was the natural consequence of group migration and of the economic and geographical forces that determined routes of settlement, was especially marked among the Germans and the Scotch-Irish . . . [T]he most striking instance of this was in Pennsylvania where German and Scotch-Irish settlers lived in communities as completely isolated from each other as they were from English-speaking communities. . . . As a result there were such constrasts in language, religion, customs, architecture and agricultural methods between the two regions that they could plausibly be compared with neighbouring states in central Europe. (Jones 1960: 49)

The establishment of a federal, decentralized system of government had a lot to do with the experience of colonial status and a desire to avoid heavy-handed rule, but it also provided a way of accommodating the diversity of the early American people. In the regions in which they were strong, each religio-ethnic group could maintain its own culture and society with little regard for what others did elsewhere.

In the eighteenth century, ethnic discord had resulted from competition between various groups of early settlers. From the 1830s one sees a different sort of conflict with the long-established settlers uniting as 'native Americans' in opposition to the great waves of Roman Catholic migration from Ireland and southern Europe. In addition to resenting the competition for economic and domestic political resources which these migrants represented, many nativists doubted their loyalty. While they owed their primary allegiance to the Pope of Rome could they be good Americans? Anti-immigrant sentiment was channelled into a number of political movements. In the 1850s, the American (or 'Know Nothing') party wanted restrictions on immigration and the curtailment of the citizenship rights of Catholics. Similar points appeared in the platform of the American Protective Association forty years later (Lipset and Raab 1978: 79–82) and in the first revival of the Ku-Klux-Klan in the 1920s. This second Klan was more concerned with Catholics than with blacks: 'there was a general feeling that the election of every additional Catholic to public office would hasten the time when our government would be turned over to a foreign Pope' (Louks 1936: 106).

The gradual diminution of religious affiliation as a major political factor clearly had complex roots and any brief listing is bound to be heavily selective. A major element was the ease of cohabitation made possible by the sheer size of America and the decentralized nature of the developing polity. Although established interests resented every loss of an increment of power and status, such losses were relatively easy to bear when change at the national level was accompanied by a large element of business as usual in many regions. Distance made accommodation easier. Conservative Protestant nativists in South Carolina might complain vociferously about the rising influence of Catholics in the cities of the North East but they were a long way away and need have little actual impact on life in Greenville or Spartanburg. The contrast with the conflict in Ireland, where Protestants and Catholics faced each other in a small land mass, is obvious.

Federal America had no choice but to tolerate and promote religious pluralism. Some colonies had legally established churches but they were not the same church. In other colonies there were vigorous advocates of toleration. Some of the Protestant dissenters tried to campaign for their own religious liberties while proposing that such rights be denied to Catholics but, in the end, the logic of their own arguments combined with the political realities of the situation to promote religious tolerance.

Generally the Catholic Church in America accepted the model of denominationalism and avoided making itself an easy target for nativist hostility.[1] As it grew, the hierarchy removed many of the powers which the laity had acquired in the early years of loose organization, but its authoritarianism was confined to its own people. On the occasions when church policy did become a matter of public policy—on the question of public schooling, for example— Catholics were sufficiently numerous and well organized in enough urban areas to force a working accommodation. It seems no coincidence that late nineteenth-century and early twentieth-century nativist movements often recruited best in those states and counties that had fewest Catholics.

Changes in the popularity of religious ideologies tend to be cyclical rather than linear. Just as the same heresies appear again and again, so organized religion stutters from enthusiasm to formalism and from stagnation to enthusiasm. Such cycles are overlaid by additional class-based processes. Evangelical and orthodox Protestantism loses first its evangelical quality and then its orthodoxy as its adherents rise in social status. But for the purposes of this discussion many of these complexities are relatively unimportant. What does matter is that, by the time of the most popular nativist movements, few members of the upper classes were conservative and evangelical in their theology. Some may still have adopted anti-immigrant positions but they were not terribly keen on active anti-Catholicism; it threatened to cause unnecessary industrial unrest and anyway no longer had a place in their religion.

Although there remained a subtle but enduring anti-Catholicism which, for example, prevented Al Smith winning the presidency in 1928, and which was only buried with the election of J. F. Kennedy

[1] Although anti-Catholicism of the sort which consumed *The Awful Disclosures of Maria Monk* was largely independent of any evidence that Catholicism was socially harmful, the hierarchy in the 1850s did make a few mistakes which antagonized more moderate Protestant opinion. Archbishop Hughes, for example, in seeking to remove control of church property from lay trustees confirmed Protestant stereotypes of Catholicism as authoritarian. For details see Billington (1964).

in 1960, by and large American national life has managed to prevent religion producing disruptive political conflict. At nation-state level there has remained a considerable amount of 'God talk'—what Bellah (1967) and others have called civil religion—but the need for this religion to avoid giving offence meant that it had so little content as to have few behavioural consequences. It is interesting to note that while God is mentioned in such key elements in political hagiography as the Declaration of Independence and the Gettysburg address, Christ is not. In so far as America was a 'nation under God', the God was not specifically Christian but was rather a Judaeo-Christian God stripped down to a deist essence. However, each religio-ethnic group was able to maintain its own religious culture at town, county, and, in some cases, state level (Lipset 1964).

FUNDAMENTALISM, ISOLATION, AND PRESSURE FROM THE CENTRE

The links between church and party show two distinct patterns. Firstly, the Protestant churches divided politically along the two lines of establishment and class. In the period from independence until almost the Civil War, the conservative Federalists were strongly supported by the Episcopalians and the Congregationalists: the churches which had been the established religion of many of the colonies and which, by virtue of their having been supported by the earliest and wealthiest settlers, had become the churches of the upper classes. The Presbyterians 'were largely a low-status Scotch-Irish immigrant group who had been rejected by the Puritans of New England' (Lipset 1969: 250). The Baptists and the Methodists had both grown rapidly by recruiting from the back country in frontier revivals. These three churches (with the Catholics) all had an interest in the disestablishment of the Congregational and Episcopalian state churches and their members were predominantly later migrants of lower status.

The early religious–political pattern in America, therefore, found the more evangelical Protestants, particularly the Baptists, the Methodists and the Presbyterians, backing the Jeffersonians-come-Democrats, along with the emerging Catholic population. The more deeply established and less evangelical Protestants, notably the Congregationalists and Episcopalians, supported the Federalists and Whigs-come-Republicans. But by the time the two-party system had been recast in its durable Republican-Democratic mould after the Civil War, the Northern Methodists, Presbyterians and Baptists were predominantly Republican. Since most American Protestants had by then become Baptists and Methodists, this realignment meant that native-born Protestants in general had become predominantly Republican. (Lipset 1969: 253)

Slavery split the Presbyterians, Methodists, and Baptists into northern and southern organizations. It is testimony to the flexible quality of religious ideology that the most actively evangelical sections of Protestantism in both the North and South took up the most extreme and opposing views on slavery. In the North, evangelicals were at the forefront of the abolition campaign; in the South they were the strongest defenders of the institution of slavery. For southern evangelicals, such as R. L. Dabney, slavery was divinely ordained:

Every man was in his place because He had set him there. Everything was as it was because He had ordained it so. Hence slavery and, indeed, everything that was, was His responsibility, not the South's. So far from being evil, it was the very essence of Right. Wrong would consist only in rebellion against it. (Cash 1954: 93)

The slavery and Civil War realignment only served to consolidate the strength of evangelical Protestantism in the South. Many northern Protestants moved from the Democrats to the Republicans via the 'Know Nothing' party and its anti-Catholicism. Originally recruiting anti-Catholics in the South as well, the 'Know Nothings' divided over slavery and the northern elements were absorbed into the Republican party, which in many north-western states took on a decidedly nativist tinge (Billington 1964: 425–30).

The involvement of northern evangelicals in the abolition campaign is important because it reminds us that evangelical Protestantism has not always been a 'reactionary' force. The theological division of evangelical and liberal only came to match the political division of conservative and liberal at the end of the nineteenth century when progressivism became thoroughly collectivist in orientation. Prior to that point, Protestants who believed that the Bible was the revealed word of God, that one had to be born again to enter the kingdom of heaven, and so on could collaborate with those followers of the higher criticism in biblical scholarship and rationalism in theology who later became 'social gospellers'. There was then no radical divide between those who thought that the world would be improved by the conversion of individuals and those who felt that the conversion of individuals was only possible once the world had been improved.

It was in response to the twin forces of the higher criticism and the social gospel that fundamentalism emerged as a distinct religious and social movement. The movement took its name from a series of pamphlets—*The Fundamentals of the Faith*—published between 1910 and 1912, which articulated a conservative reaction to these two powerful tendencies in the major Protestant denominations (Marsden 1980). The fundamentalists were those who refused to accept the new thinking. They continued to insist that the Bible was the word of God, that miracles really happened, that unless one had experienced religious conversion one was destined for an actual hell, that Genesis 1–12 provided an accurate account of the origins of life and species, and so on. In the first stage of the reaction there were fundamentalists in all the denominations (although there were proportionately fewer in those denominations—the Episcopalian and the Congregationalist—which recruited primarily from the upper classes). As it became clear that the majority of Protestants would not be diverted from their apostate ways, the fundamentalists withdrew to form their own conservative denominations.

They did not stop at church formation. In those areas, such as most of the South, in which they were strong, their culture dominated a wide range of social institutions; where they did not have such predominance by virtue of weight of numbers in the general population, they created their own institutions. They built schools, colleges, and universities. They produced their own papers and magazines. They built radio stations and, when the technology presented itself, they were quick to appreciate the potential of television. They made programmes and paid companies to air them. Finally they created their own television networks.

In sociological terms, fundamentalists were working to create social institutions which would permit them to reproduce their own culture sheltered from modernizing influences. The problem was that such boundary-maintaining activity depended on a weak (or benign) federal government which would permit the regions (and pockets of fundamentalists more centrally located) to go their own way. Although it was clearly the intention of the drafters of the Constitution that the central government should be subordinate to the states, the whole thrust of modernization has been towards an active and powerful centre. An early and traumatic example of the centre's unwillingness to permit the peripheries to follow their own interests was the Civil War (which actually produced far less change in the South than had been intended). By the 1950s and 1960s, it was not the Army but the Supreme Court and the Congress that were imposing liberal and cosmopolitan values on the South. Without multiplying examples one can characterize the last hundred years as a period in which regional autonomy has gradually been eroded by central government intervention on issues ranging from spending on public works to the drawing of constituency boundaries.

As is generally the case with major wars, government intervention increased markedly with American involvement in two world wars. And it continued to increase. In 1976, there were seventy-seven federal regulatory bodies and fifty of them had been created since 1960 (Janovitz 1978: 368). Federal funding has become an ever-larger element in state spending. At the same time the distinctiveness of regional culture has been threatened by the growth of national corporations, the population movements caused by four wars, and the gradual concentration of the media.[2] Putting it simply, the 'Bible belt' was penetrated by cosmopolitan culture. The South had oil, cheap non-union labour, and sun, the heavy industries of the North were in decline, and people were moving into the South.

If the gradual ending of the isolation which had helped preserve their world was one source of concern to fundamentalists, another was the increasing permissiveness of the culture from which they can no longer remain isolated. Those commentators who wish to stress the resource mobilization view of the genesis of social movements will correctly observe that fundamentalists have always been at odds with elements of the setting in which they find themselves; theirs is a prophetic religion. The danger is that such an observation may be combined with an abnormal psychology view of conservative Protestants to produce the idea that the fundamentalist critique of the world is unconnected with the substantive

[2] Although American media are considerably more open and diffuse than their British counterparts, there has been a gradual concentration of print, radio, and television journalism as major corporations have bought out small outlets and as individual outlets have come to rely more and more on centrally produced and syndicated copy.

reality of the world they criticize. We should not forget that the America of the late 1960s and 1970s was a very different place from Eisenhower's America. Where once two-thirds of the states' legislatures were willing to vote for prohibition, there was open campaigning for the legalization of marijuana. Where once divorce caused considerable social stigma, there were single parent, lesbian, and male homosexual 'families'.

The increasing affluence which made the 'permissive society' possible was also enjoyed by conservative Protestants. They could afford to indulge themselves. The religion which had previously made a virtue of their straitened circumstances by promoting asceticism now, for some, seemed like a strait-jacket of petty constraints. As fundamentalists tried to evolve new standards and new responses, the much more dramatic changes in patterns of consumption and expression which they heard about on their radios and then saw on their televisions seemed like the work of the Devil. Because they now had the money and leisure to be able to indulge themselves in dancing, some tentatively tried it while the majority read 'Twenty Reasons Why I will Not Dance'. The American journalist Frances Fitzgerald (1986) appreciates the point when, in a collection of her writings, she places an essay on the gay rights movement in San Francisco next to a report on Falwell's Moral Majority and Liberty Baptist University. The increasing permissiveness of American culture was a threat to the fundamentalist way of life, not only because it rejected the principles of control and discipline which fundamentalists held dear, but also because it had roots in the increasing affluence which many fundamentalists shared and which presented them with new temptations. While some compromised and moved from a classic fundamentalist position to a more moderate 'neo-evangelical' stance, others fought back.

The other great change in American life was the promotion and acceptance of group rights. Blacks, women, and then homosexuals claimed rights as members of previously disadvantaged groups. Such claims offended conservative Protestants for two reasons. Intellectually they were abhorrent because they ran counter to the individualism of conservative Protestantism. Since the demise of Calvinism, the conservative American Protestant tradition has been consistently Arminian. *Individuals*, not groups, get saved. Collectivities cannot be 'born again'. The second reason for rejecting the claims of various 'minorities' was perhaps less clear but equally potent. The success of any such claims would likely result in major changes in power, influence, and authority.

To return to a theme of the first chapter, to say this is not to endorse the old status defence or status anxiety theories of the origins of social movements. There was certainly an element of status displacement involved in the civil rights movement in that, if blacks had succeeded in improving their socio-economic position, it would have produced a relative weakening of the position of WASPs, but what appears to have caused far more concern was the displacement of those beliefs, symbols, and patterns of behaviour which fundamentalists held dear. It may be that, for some, the belief that blacks were inferior, that races should not mix, or that women should obey their husbands, was a convenient endorsement of social and political arrangements which they found congenial for more material reasons. But ideological legitimations of patterns of action become embedded in consciousness and they become important resources in their own right. Challenges to those ideas were potent threats.

Further insight into the conditions which created a market for the new Christian right can be gained from a consideration of the social proximity (as distinct from just geographical proximity) of what was construed as the enemy. The old Christian right was concerned with the threat from 'world communism'. In the 1940s and 1950s, many fundamentalists were active supporters of anti-communist movements. Some, like Billy James Hargis and Carl McIntire, foreshadowed such NCR television evangelists as Jerry Falwell by using their fundamentalist radio ministries to promote anti-communist organizations. Even more moderate fundamentalists, such as the writers of *Moody Monthly*, gave considerable attention to the need to oppose the red peril. Contemporary conservative Protestants remain anti-communist.[3] But communism was somewhere else; there was very little of it in America, on the doorstep. It required considerable ideological work to see reds under the bed. Many were prepared to do that work but it took less effort to see the widespread permissiveness of post-Camelot America as a

[3] A large confidential survey of potential NCR supporters conducted for Christian Voice shows that anti-communism remains one of the most potent unifying values. I am grateful to Gary Jarmin of Christian Voice for allowing me to see some of the first analyses of this data.

threat. The changes which concerned the fundamentalists of Falwell's generation were occurring in America and even when they were not happening in Greenville, South Carolina, they were visible on the television screen.

Far more could be said about the reshaping of American society and culture, and about the impact of those changes on fundamentalists. It is enough at this point to establish that there were actual changes in the environment, which concerned fundamentalists, and to which they *could* respond by active opposition. I put it like this to make it absolutely clear that I do not suppose some automatic or immediate connection between changes in the environment and any particular response. Resource mobilization theorists are quite right to stress the need for mediation. Collective action is not simply the automatic response of lots of isolated individuals to circumstances which they find worrisome or troublesome. One needs the additional element of 'consciousness-raising'. People have to be persuaded (a) that they share common concerns, (b) that there is a known and identifiable cause for what concerns them, and (c) that collective action of a particular kind will remedy the situation. That is, interpretative work is required to move people from their unfocused and unorganized sense of being ill at ease with some aspect of the world to a shared understanding of what is wrong and what can be done about it. This interpretative work will be the focus of the next chapter but before turning to that, one particular source of threats to conservative Protestantism—the courts—will be discussed.

THE COURTS AND RELIGION

The court system is an important agent of increased centralization in American life. As their part in the extension of civil rights demonstrated, the courts have been a major source of cosmopolitan and modernizing values. They have also been heavily involved in disputes about the proper place of religion. In this section I will briefly describe the structure of the American court system, explain why it is more politically active than the British judiciary, and review the recent history of those judicial decisions which have had an impact on conservative Protestants.

There are two American court systems: state and federal. Federal courts deal with offences against federal law, disputes between states and citizens of different states, and matters of constitutionality. State courts deal with most other things.

States differ in the ways in which they acquire their judges and justices (the members of the superior or supreme court) but many elect the judiciary. Given the low turn-out rates common in American elections and the general lack of interest in the races for positions at the bottom of a long list of elective offices, many places in the state judiciary are 'political' in the sense that office can be secured through the patronage of the local party machine. There is also an element of political patronage in the federal judiciary. Federal judges are nominated by the President for the approval of the Senate, but the President's initial recommendations are often suggested by the political leaders of the district which has the vacancy. However, while some federal judges will have been nominated because they have strong local ties and loyalties, the federal court structure (and, to a lesser extent, that of the state courts) does promote cosmopolitan values because the appeal system means that the decisions of large numbers of lower courts, staffed by judges drawn from the local bourgeoisie, are reviewed by a small number of senior courts. It is centripetal.

If a case comes under federal jurisdiction or, having originated in a state court system, is adjudged to involve a principle of constitutionality, it may be appealed to the Supreme Court. Given the social and educational background of the various levels of the judiciary, it is inevitable that the higher courts should be more liberal and cosmopolitan than the lower courts. The particularistic decisions of the legal system at county and state level are judged by the universalistic standards of the higher courts.

> The centralization of authority flowed from the appeals process itself. When representative democracy, the Republic, was deliberately unresponsive, as it was to black Americans, self-interested individuals found a mechanism to begin constructing a new political system, judicial appeals became the politics of the day and the appellate process became the process of governing. There were national answers to local questions, then there were fewer and fewer local questions. (Reeves 1982: 151)

The centripetal tendency has been amplified by the activist attitude adopted by many senior judges and justices of the post-Second World War era. Richard Thigpen, the vice-chancellor of the University of Alabama, argues that the courts 'are the only instrument of government which regularly and consistently and visibly get things done' (in Reeves 1982: 150). As Richard Neely (1981), the Chief Justice of the West Virginia Court of Appeals, has argued in describing his own judgements, the considerable extent to which elected legislators are hamstrung by the frequency of elections has created stalemate. Politicians are loath to act, even when they privately recognize that some issues must be addressed. Action, especially on behalf of the poor and the weak (who are disproportionately non-voters), offends more voters than it pleases and the 'general good' of the community does not vote. Hence serving that good is rarely a vote winner. Although many judges are elected, many are appointed, many are elected for life, and even those who face re-election have considerably longer terms than most politicians. Although they are more beholden to the public (as electorate) than are their British counterparts, American judges and justices are far less beholden than are American politicians. Hence it is no accident that most progressive social change—in the area of race relations, for example—has been initiated and promoted by judicial decisions rather than by actions of the legislature.

CHURCH AND STATE

Given the importance of court decisions in the history of the NCR and the considerable confusion that surrounds the present interpretations of the doctrine of the separation of church and state, it is worth spending some time clarifying the past and present positions. Relationships between government and the churches become a court matter because of the First Amendment to the Bill of Rights, which states: 'Congress shall make no law respecting an establishment of religion, or prohibiting the free exercise thereof . . .' Conservatives who believe that the Supreme Court has gone too far in a secularist direction in interpreting this brief injunction construe 'establishment' in a narrow way; they interpret the Amendment to mean only that the government should not support one church or sect over another. In contrast, liberals point to the speeches which James Madison made in favour of his Amendment to argue for a broader view. It would certainly be curious if the supporters of the Amendment had only meant to prohibit the elevation of one church over others because, by the time the Amendment was being debated in 1789, only five states still had legal establishments and they were all of the 'multiple' kind, of which South Carolina's is a good example. South Carolina's Constitution declared: 'The Christian Protestant religion shall be deemed, and is hereby constituted and declared to be, the established religion of this State. . . . All denominations of Christian Protestants in this State . . . shall enjoy equal religious and civil privileges' (in Butts 1986: 17). Although churches were supported by local taxes, tax-payers could specify which organization received their support. For Madison and others to be arguing against 'establishment of religion' in a time when singular establishment had already given way to the multiple form (and no one was arguing for the change to be reversed) seems to suggest that they were opposed to government support for religion *per se*. The second part of the Amendment, the part that conservative Christians stress, seems obvious in its intentions. While the government should not promote religion, it should also do nothing to abridge the individual's right to the free exercise of his or her religion. In addition to the problem of interpreting each clause, there is, of course, the difficulty of reconciling them.

The drift from singular to plural establishments was a sensible reaction to the increasing pluralism of American religion. Liberals like Madison and Jefferson seem to have wanted to take this even further. When Jefferson was promoting the foundation of free public schooling in Virginia, he insisted that 'no religious reading, instruction or exercise, shall be prescribed or practised inconsistent with the tenets of any religious sect or denomination', which effectively meant very little could be practised. Jefferson thought it fit 'to leave every sect to provide, as they think fittest, the means of further instruction in their own peculiar tenets' (in Wood 1972: 400). It was Jefferson who first coined the phrase 'wall of separation' to describe what he saw as the correct relationship between church and state.

Most public schools were not as rigorous in avoiding the violation of 'the tenets of any particular sect . . .' as Jefferson might have wished. However, as long as the majority of the people were Protestant, there was little problem in retaining Bible reading and public prayer.

For more than a century the courts were able to sustain a principled opposition to any formal religious establishment because an informal establishment already existed . . . there was no question that ours was a WASPish sort of hive until late in the nineteenth century. (Demerath and Williams 1987: 81)

The problem came with the gradual 'deghettoization' of the growing population of Roman Catholics. Catholics found the public schools offensive on two grounds. Ironically, the first was that, like many Protestants, they wanted education suffused with religion. They just wanted that religion to be Roman Catholic. The second was that, while the Bible readings and public prayers might not have offended any Protestant sect, they did offend Catholics. The Bible commonly used was the 'Protestant' King James version and the prayers were Protestant.

Roman Catholics established their own separate school system. Initially it was conservative Protestants (now keen to reintroduce religion to schools) who most aggressively campaigned for 'the separation of church and state' because in the mid-nineteenth century that slogan translated as 'no public funds for Catholic schools'. The arrival in the 1880s and 1890s of large numbers of Eastern Orthodox and Jewish, as well as Roman Catholic, immigrants, simply increased the tensions and led many states to ban religious exercises in public schools.

Given the tendency of fundamentalists to blame the present prohibition on religion in public schools on the Supreme Court, it is worth noting that many states struck separatist positions much more rigorous than anything the Court recommended. At the time of the historic *Schempp* v. *Murray* decision of 1963, eleven states explicitly prohibited Bible reading and religious exercises in schools; nineteen states permitted Bible reading in that they did not have statutes actually prohibiting it; six states had statutes permitting Bible reading; and only eleven states had laws requiring Bible reading. With the exception of Idaho, all of these were in the South or along the eastern seaboard.

Furthermore, the presence which religious exercises had in schools by 1948 was almost entirely the product of recent legislative activity. With only one exception, all the statutes which required Bible reading in public schools had been enacted after 1913. Far from following the obvious (for an industrializing society) trajectory of increased secularization, twentieth-century America has seen an increase in church membership. This appears to have combined with a growing appreciation among Protestant denominations that no one of them would dominate the others, to produce a demand for *more* religion in schools. The Catholic Church's parochial schools relieved Protestants of the need to consider Romanists. However, the return of religion to the schools was not accompanied by any reduction in the fragmentation of America's religious culture and, inevitably, the Supreme Court was called upon to judge the constitutionality of various laws and practices.

In its judgement on the 1948 *McCollum* case, the Court declared that 'released time'— setting aside a portion of each day for religious instruction by representatives of various faiths—was unconstitutional, even though attendance was purely voluntary. By 8 to 1, the justices supported the Madison view that 'no establishment' meant more than that no one denomination should be given preferential treatment. To quote from the majority judgement:

> . . . the First Amendment rests upon the premise that both religion and government can best work to achieve their lofty aims if each is left free from the other within its respective sphere. Or, as we said in the Everson case, the First Amendment has erected a wall between Church and State which must be kept high and impregnable. (in Wood 1972: 405)

In the 1952 *Zorach* case, the Court repeated and strengthened its decision and in 1962, in *Engel* v. *Vitale*, the Court said that it was immaterial whether a state-sponsored prayer programme was non-denominational, optional, or involved the use of tax funds. Prayer was a religious act and therefore could not be sponsored by the state.

The 1963 *Schempp* v. *Murray* case asked the Court to rule on the constitutionality of the by then widespread practice of Bible readings in public schools. The Court repeated the position of earlier judgements that the First Amendment was intended not just to outlaw the establishment of any one church but also to create a complete separation of spheres between church and state. While the study of the Bible or religion were not outlawed, provided they were studied objectively, devotional Bible reading and prayer recitation were unconstitutional because they 'are religious exercises, required by the States in violation of the command of the First

Amendment that the Government maintain strict neutrality, neither aiding nor opposing religion' (in Wood 1972: 406). As in *McCollum*, the *Schempp* decision was nearly unanimous: 8 to 1.

In a series of judgements in the 1970s, the Court elaborated on its tests for unconstitutionality. Any law will be struck down if it fails to pass the threefold test of purpose, consequence, and entanglement. If the purpose is primarily to advance the cause of religion, then it is impermissible. Irrespective of 'purpose', if it can be reasonably inferred that a consequence of the law is the advancement of religion, then it is unconstitutional. Finally, all such judgements must bear in mind the need to avoid excessive entanglement of government in the affairs of religion.

THREATS TO THE SUBCULTURE

Conservative Protestants have been greatly offended by the Supreme Court's commitment to the Madison–Jefferson wall of separation position. In their umbrage at what they see as the banning of school prayer, they have usually overlooked the Court's record in protecting the individual's right to free exercise and in expanding the definition of what counts as a religion deserving such protection. That their right to say whatever prayers they liked at home has been protected made little impression on the conservative Protestants who saw a ban on prayer *in schools* as being a ban on prayer *per se*.

This selective attention is not surprising. After all, free exercise cases have usually concerned the rights of despised minorities such as the Jehovah's Witnesses. Although I will later argue that the new Christian right has been most successful when it has presented itself as the representative of a persecuted minority interest group, the self-image of fundamentalists is that of a 'Moral Majority'. Conservative Christians were deeply offended by the Court's clear refusal to take sides. To judge all religions as equally worthy of protection and to favour none was, to true believers, the same as saying that their religion was false. Conservative Protestants were not impressed with the right to say prayers at home with their own families, or in church with other believers; they wanted the right to have their prayers said in public places, by large numbers of people outside their own church fellowship. To put it bluntly, they wished to see their culture in a position of pre-eminence and the Court's decisions thwarted that desire.

When one's values are not given pride of place in a society's culture, it is always possible to construct a smaller sub-society with its own subculture. This is generally the Roman Catholic response to minority status and fundamentalists had already begun to move in this direction in the 1920s when the mainstream churches became more liberal and ecumenical. The amount of social institution building they had to do was always inversely related to their power and influence. In those areas where they remained strong, they created Bible colleges and universities but only the most extreme fundamentalists felt that the public schools in rural and small town South Carolina, for example, were so secular as to require alternative Christian schools. Although many schools in fundamentalist areas continued (and still continue) to defy the Supreme Court and begin the day with Bible readings and prayers, the increasingly secularist position advanced by the judiciary has led many fundamentalists to abandon the public schools altogether.

The independent Christian school movement began slowly but gathered pace in the 1970s. Whereas overall school enrolment declined by 13.6 per cent between 1970 and 1980, the number of independent Christian schools grew by 95 per cent. They now have around 2.5 million students in between 17,000 and 18,000 academies (Hunter 1987: 6). Critics of fundamentalist schools explain their growth as a response to desegregation rather than to secularity. Parents who were unwilling to have their children educated with blacks opted out of the public schools. The vast majority of conservative Protestant school founders deny racist intent and instead assert that it was the generally liberal socio-moral climate of the public schools which they found offensive.

Withdrawal did not so much end church–state arguments as shift their focus. Liberals had opposed the teaching of religion in schools on the grounds that it amounted to the state conferring additional legitimacy on one particular religious ideology (an argument which they felt remained relevant even when the religious ideology was simply a bland Protestantism). But in addition to the legitimacy question there was also the issue which had been raised in the previous century over the question of state funding for parochial schools. Catholic parents had then argued that they paid for the education of their children twice; once in the taxes which supported a public school system they did not use, and again in fees and donations to church schools. Most Protestants firmly opposed government aid to parochial schools and the Supreme Court has generally regarded any such aid as unconstitutional.[4] In the last decade, the debates have moved to question the tax-exempt status of independent Christian schools.

In America, as in most modern states, religious organizations are exempt from taxation. Why this should be the case is itself an important point of disagreement. One 'functionalist' view is that organizations such as churches are exempt from taxation because they perform good works. Even those who would not define the propagation of religion as a good work can accept this by concentrating on the social service aspect of most religious organizations. Others object to the idea that tax-exemption is a form of public subsidy and argue instead that churches are not taxed because they do not make profits and because their members already pay taxes as wage-earners and consumers. That is, religious organizations are not exempt; they are simply not properly part of the tax-base.

The Internal Revenue Service (hereafter IRS) has taken the first line of argument. In the middle 1970s it began to challenge the charitable (and hence tax-exempt) status of fundamentalist academies. The IRS argued that charitable status was effectively a public subvention and was only merited by institutions which demonstrably served the public good. Although a clear positive definition of the public good was not presented, the IRS argued that certain activities could be clearly seen to be at odds with public policy. Through a series of judgements and actions (and failure to act), the Court and Congress had established that racial integration was public policy. Hence any organization which was segregationist could not be a charity. Most of the fundamentalist academies could argue in their defence that they were not positively segregationist in that they did not prevent blacks from enrolling;

[4] The present legal situation is extremely confused. Courts have been prepared to permit state support for 'peripheral' activities, such as transport to and from denominational schools or the purchase of secular testing materials. The most comprehensive accessible source of summaries of court decisions is *The Schools and the Courts*, a quarterly published by School Administration Publications of Asheville, North Carolina.

that they had very few black students was the result of black parental choice.

In 1980 the newly elected Reagan Administration tried to stop the IRS continuing its case against Goldsboro Christian Schools, North Carolina, and Bob Jones University of Greenville, South Carolina. Liberal reaction was so strong that the Administration backed down and allowed the action to proceed to the Supreme Court. Although the case against Goldsboro Christian Schools was clear in that it was avowedly segregationist, the matter of Bob Jones University was far less so. Initially segregationist, BJU had bent with the wind and enrolled non-white students. However, it prohibited interracial dating. BJU fundamentalists believe that interracial marriage is unbiblical. As dating is only acceptable as a preliminary to marriage, interracial dating should not be encouraged. It is in the nature of BJU that activities which are undesirable should be straightforwardly prohibited.

The Supreme Court was remarkably united in its rejection of the BJU case. Only Justice Rehnquist (who was promoted to Chief Justice by Reagan in 1986) completely dissented from endorsement of the IRS action. The Court accepted that charitable status was dependent on accordance with public policy and that Congress had made its will in such matters quite clear.[5]

In other cases, Christian schools have been closed for refusal to submit to state licensing and fundamentalist pastors have been imprisoned. A Nebraska state law required that the state ensure that all children receive a satisfactory education. Although many independent Christian schools would have met the required standards, a number refused to submit to state approval and were ordered to close. In his judgement on the constitutionality of the laws requiring state approval, the Senior District judge said that the state's compelling interest to ensure that students were properly taught overrode parents' religious convictions.

The Supreme Court does not judge general issues in the abstract. It can only hear particular cases which have passed through the lower courts, which raise fundamental constitutional issues, and in which the plaintiff possesses certain characteristics of 'good standing'. These limits on the nature of the cases the Court hears means that there is always a problem in inferring general principles

[5] The Supreme Court judgements are 81–3 and 81–1, 4 May 1983.

from the Court's decisions. As will also be clear, there is often a tension between the establishment principle and the free exercise clause. While some segregationist schools have had their tax-exempt status revoked (on the establishment principle that tax-exemption is a form of state support), they have been allowed to operate (on the grounds of free exercise). To conservative Christians who wish to enrol their children in such schools, the Court's position is further confirmation of a deliberately secularist policy. As they tell it, firstly religion was taken out of public schools and then the right to educate their children privately was eroded by administrative action which raised the costs of such withdrawal from the public school system.

To summarize, the history of Supreme Court judgements on church–state relations can be seen as a gradual response to increasing religious pluralism. At the same time as the Court has been willing to extend the protection of the free exercise clause to smaller more deviant religious groups, it has sought to erect and maintain the clear wall of separation between church and state which Jefferson advocated in the last quarter of the eighteenth century. To conservative Protestants, the decisions of the Court are simply further examples of the increasing power of secularists. The importance of this story for understanding the rise of the new Christian right is that decisions about the constitutionality of prayer and Bible reading in schools (and about related issues which will be discussed later) have enormous symbolic value. Although the justices have often been at pains to state that they have nothing against religion in its proper place, they have made it clear that its proper place is the home and the family, not the public arena. This is sufficient offence to fundamentalist sensibilities. After all, their refusal to go down the liberal road is based on rejection of the idea that religion can be confined to a small part of one's life-world. But the offence is made even greater by the gradual expansion of what the modern state regards as the public arena.

CONCLUSION

The relative isolation of the geographical peripheries of America and the decentralized polity which permitted subcultures to form sub-societies explains why America retains to this day a larger proportion of conservative Protestants than other modern industrial societies. The increased government intervention in many areas of life, the increased permissiveness of cosmopolitan culture, and the increased secularity of the interpretation of the constitutional directive to separate church and state explain why a large number of conservative Protestants should have become increasingly concerned about the future of their culture. To a limited extent such concern could have been translated into further isolationism. Even if state interventionism continued, conservative Protestants could have become fatalistic and derived an unholy pleasure from the thought that the rest of America was heading for damnation while their salvation was assured. We thus need some explanation for the choice of active opposition to secularizing liberal culture and such an explanation can be found: (a) in a number of broad socio-cultural changes which were enhancing the power and influence of the South; and (b) in various elements of the political structure which promised some hope of success to the aspiring activists.

The analytical significance of this second point is simple. In formulating definitions of situations, reactions, and possible plans of action, people imaginatively anticipate the consequences of their actions. A strong suspicion that a particular plan will not succeed often feeds back to affect not only commitment to the planned action but the definition of the situation itself. Even once a plan of action is taken up, it is often adopted tentatively so that consideration of the consequences of initial acts influences guiding schemes and definitions. Although at first sight this proposition might seem to be violating the irreversibility of the passage of time, expected outcomes and outcomes themselves are things which people consider when defining situations as problematic and as calling for action. Thus the structural elements which explained success (or failure) will be relevant to any explanation of why people were willing to be mobilized for a certain course of action. The appreciation that the American political system, the organization of the mass media, and other things which will be discussed later in explaining the mobilization and partial success of the new Christian right might be conducive to such a movement is itself part of the explanation of the genesis of the movement. However, for simplicity of presentation, these considerations will be discussed after an account of the mobilization of the new Christian right and they will be presented in answer to the question of why Britain has not witnessed a similar movement.

To return to 'the rise of the South', it is worth noting a truism of political and social movements. Although most movements are directed to the solution of some problem, the remedy of some grievance, or the alleviation of some hardship, the genesis of many movements follows periods of *improvement* in the circumstances of those mobilized. Improvement not only increases expectations (and hence a sense of relative deprivation); it also improves morale and makes the oppressed more optimistic about the possibility of improving their conditions further. A large number of social and economic changes have raised the status of the South relative to the rest of America. Race relations have ceased to be a major source of stigma, not because southern blacks have been thoroughly integrated, but because the end to formal segregation in the South and increasing awareness of the economically depressed condition of blacks in the North have reduced the ability of northern whites to claim superiority on the grounds of better race relations. Although southern whites remain less sympathetic to integration than whites in the North, positions on the worth of the principle of integration have moved close together (Schuman, Steeh, and Bobo 1985: ch. 3). At the same time as southern whites have come to accept the principle of public integration, northern whites have become less sympathetic to government intervention to turn that principle into reality.

Changes in the economy—the decline of the old heavy industries and the increasing importance of oil—have caused a shift of power from the North to the 'sunbelt': the old South plus the South West.[6] The low levels of unionization in the labour force, the absence of local government regulation, and the low levels of local taxation have all made the sunbelt attractive to many industries and the weather has made the South attractive to wealthy retired people. In addition, many high technology and aerospace enterprises associ-

[6] The term 'sunbelt' was introduced into political discourse by Kevin Phillips in 1968 to refer to the area from 'the Charleston-Savannah-Jacksonville coastal strip to California's urban South'. Later he cited the 37th parallel, 'a line a bit north of the North Carolina–Tennessee border, to the west following the Oklahoma, New Mexico and Arizona boundaries, and then cutting through Nevada and California'. For a discussion of the sunbelt and the 'balkanization' of America, see Phillips (1982: ch. 7).

ated with the military have located in the sunbelt. Although it is always dangerous to offer one figure or one election as a symptom of a major trend, the election of Jimmy Carter, ex-Governor of Georgia, as President in 1976, the first southerner to win such high office since the Civil War, can be seen as an important indicator of a more general rise of the South, as can his defeat by Ronald Reagan, the ex-Governor of California.

CONSERVATIVE GROWTH AND TELEVANGELISM

Another important factor in the increasing self-confidence of American conservative Protestants has been the growth in their numbers and presence. Until the 1960s all the major denominations were showing an increase in their membership. 'Since 1965, membership in liberal denominations has *declined* at an average five-year rate of 4.6 percent. By contrast, evangelical denominations have *increased* their membership at an average five-year rate of 8 percent' (Hunter 1987: 6). The most successful sections of Protestantism have been the Southern Baptist Convention and the independent fundamentalist Baptist groups.

The reality of genuine growth has been further amplified by the fundamentalists' success in using television. The role of 'televangelism' (to use Hadden and Swann's term) in the mobilization of the new Christian right will be discussed in the next chapter. Here I want simply to point to the considerable morale-boosting effect of the high profile which fundamentalist preachers acquired in the late 1960s and 1970s. Although there are considerable technical problems in measuring accurately the size of television audiences, the best recent estimate concludes that 'four in ten households tuned to one or more of these religious television programs . . .' (Clark and Virts 1985: 21). Although the more overtly political preachers, such as Jerry Falwell, were not the market leaders (Hadden and Swann 1981: 55–65) they were well to the fore and even the shows of avowedly non-political evangelists such as Oral Roberts and Rex Humbard have played a major role in popularizing and legitimizing conservative Protestantism.

Fundamentalists felt threatened and offended by the values and actions of liberal America but at the same time they saw their own churches expanding, they saw new churches being planted, and increasingly they saw their kind of preacher on the television. With an almost childlike naïveté, they revelled in the evidences of their success. In order to grasp the scale of organizational growth and development in parts of the fundamentalist milieu, it is worth describing one operation—Falwell's Thomas Road Baptist Church. Falwell began his ministry in a disused soda bottling plant in Lynchburg, Virginia, in 1956 with a handful of members. In 1964 a new 1,000 seater auditorium was constructed. Three years later ground was broken for a new church with three times the capacity. With current membership above 18,000 and some sixty associate pastors being employed, even both halls are not big enough and special events are held in the local sports arena. From the first, Falwell had a vision of creating an all-encompassing world. First, a Christian school was added, and then Liberty Baptist College. By 1976 the College had over 1,500 students. In 1985, with accreditation from the Southern Association of Colleges and Schools, the institution was renamed Liberty University and its range of courses markedly expanded beyond those directed to training future pastors. In 1959 a home for alcoholics had been established. Later, when abortion became a major concern of conservative Christians, a nursing and adoption home was founded to provide a positive alternative to abortion.

The church bought an island in the middle of the James River, renovated the buildings, and opened a Christian summer camp for young children. Senior citizens clubs were organized. Large numbers of buses were bought to ferry to church children whose parents were not church members.

The services in Thomas Road Baptist Church are recorded and broadcast on (in 1981) 392 television and 600 radio stations. With a total income in excess of $60 million per annum, Falwell's ministry is now one of the major employers in Lynchburg. And Thomas Road Baptist Church is by no means unique. On only a slightly smaller scale, 'Bible Baptism' has shown similar growth in many American cities.

Conservative Protestants draw considerable comfort from the symbols of material success which their pastors acquired. Liberal critics have published books with snide titles such as *Praise the Lord and Pass the Contributions* (Bestic 1971) but to fundamentalists it is a source of pride that they have moved from clapboard shacks with a radio antenna to cathedrals purpose built for television. On my first visit to Lynchburg, one of Falwell's assistant pastors took me past Falwell's house and, with no sense of irony, asked me if any of our pastors had a house like that!

At the same time as fundamentalist denominations were growing, parts of the ideology of born again Christianity were acquiring a new popularity with a series of celebrity conversions. There is no doubt that the degree of renunciation associated with the conversion experience declined as claims to it became more common outside the traditional fundamentalist constituencies. None the less, the spread of the rhetoric (even when barely accompanied by the ideology) encouraged fundamentalists to become more assertive.

My argument is that the market for the new Christian right was created by the interaction of a number of trends. On the one hand, an increasingly liberal state, committed to promoting interests (such as those of blacks, women, and homosexuals) which fundamentalists opposed, was intervening more and more in the sub-societies and subcultures of conservative Protestant America. On the other, the rise of the South and the stronger profile of evangelical religion gave conservative Protestants reasons to feel both that they were not getting their due and that their due could be got if they organized to claim it. The next chapter will consider the ways in which the market created by these conditions was exploited.

3

Exodus: Mobilization

IF the first part of the explanation of the genesis of any social movement concerns the market for such a movement, the second part must examine the ways in which the movement was initiated and led. This chapter will describe the mechanics of the mobilization of the new Christian right. But, as a preliminary, it is worth making a brief aside to consider the possible motives of the leaders.

The two end points of the range of possible motives for involvement in any social movement are self-interest and altruism. To offer a simple example, blacks involved in the American civil rights movement of the 1960s were acting to promote their own interests and the interests of other blacks. Although the whites who joined them were presumably acting in pursuit of some abstract goals which they held dear, they were not promoting interests that could in any obvious and immediate way be described as 'their own'. If one describes that sort of involvement as altruistic, it seems clear that there is a quite different sort of technically disinterested motivation. Certain groups may not actually desire the goals of a social movement and yet still support it because they hope that the movement will inadvertently serve their quite different interests. That is, cynical and manipulative involvement is a possibility.

This type of involvement is raised here because, although no serious scholar has suggested it, much public wisdom on the new Christian right sees it as the product of the Machiavellian intentions of 'secular' conservatives; people who did not themselves support the NCR's socio-moral positions but who saw political advantage in the encouragement of a right-wing movement (see, for example, Weeks 1987). As I will make clear later, such wisdom is increasingly shared by fundamentalists. Although such an argument has not yet been clearly formulated and proposed, it is worth considering.

CLASS, POLITICS, AND RELIGION IN MODERN SOCIETY

This subheading is borrowed from the title of an essay by Lipset (1969: ch. 5) which examines a problem common to all right-wing parties in modern democracies—their theoretical minority position—and possible responses to it.

As T. H. Marshall argued in his seminal essay on citizenship, the extension of the franchise to the mass of the population is normally accompanied by a shift to the left (Turner 1986). The expectation of such an outcome predisposes conservatives to argue against franchise extension. If people choose their politics on the basis of rational economic interest, conservative parties will probably not win democratic elections. Conservatives may try to exclude the lower classes but to do so is to threaten the stability of the democratic system because its legitimacy rests on (a) the shared belief that all significant actors have an equal access to power and (b) the expectation of some degree of rotation of political power.

That conservatives do win elections represents an obvious deviation from a purely class model of political choice, and Lipset suggests two sources of deviation. The first derives from the workings of the stratification system, which may lead lower strata to accept the values and orientations of the élites, and/or may divide members of the lower strata against themselves. The second source of deviation is the promotion of non-economic values as politically relevant.

The first source is not central to my concerns and can be passed over with only the recognition that Lipset is right to point to elements of stratification systems as being, in effect, self-legitimating, in that people who even minimally accept the dominant views of how their position can be improved will also tend to accept those ideologies which justify the existing pattern of stratification in the first place. Furthermore, even if one does not accept that the ruling ideas of any age are the ideas of the ruling class, it still seems unexceptionable that élites are in a better position than lower strata to disseminate their view of the world.

It is the second source of deviation from a class-based model of politics which concerns me. As Lipset puts it:

> Conservative parties in a political democracy usually seek to reduce the saliency of class as a basis for party controversy in order to win more

lower-class votes. These efforts may take the form of . . . stressing non-economic bases of cleavage such as religious or ethnic differences or issues of morality. (1969: 165)

However, before one accepts the assumption of some Marxists that the ruling class will stimulate religious divisions in order to keep the working class divided, it is worth remembering that most anti-Catholic movements have been populist and have received minimal support from the upper classes.[1] It could, of course, be argued that the ruling classes are too clever to get their hands dirty with such campaigns when they could gain all the benefits by standing back and letting others act in their interests. But it is equally likely that the reluctance of élites to become involved in anti-Catholicism is a result of an appreciation that the benefits of a divide-and-rule strategy, in terms of preventing the development of a working-class consciousness, are likely to be outweighed by disadvantages. The basic logic of modern capitalist development is towards the privatization of most forms of particularism. Especially when a democratic polity permits some power to religious minority groups, capitalists will seek to maximize return and predictability by promoting universalism in the economy and in those areas of public life most closely associated with it.

This is not to say that particular sections of the bourgeoisie will not act in a religiously particularistic fashion. They may well do so (as the Protestant bourgeoisie has done in Northern Ireland) but there is no reason to suppose that they do so for cynical and opportunist reasons. It seems quite clear that members of the Ulster Protestant bourgeoisie 'played the Orange card' because they themselves shared the cultural values of the Protestant section of the working class. In the case of the NCR, it would not take much ingenuity to imagine how secular conservative interests might be served by a mobilization of socio-moral conservatives, and one can identify promoters of the NCR behaving with varying degrees of cynicism. But it remains the case that those people most active in promoting the politics of socio-moral issues are themselves committed to that agenda, independent of other interests which might be served. Or, to reverse the account, as with previous populist and

[1] On populism, see Canovan (1981) and Lipset and Raab (1978). On the problems of recruiting middle-class support for anti-Catholicism, see Billington (1964: 247–300).

nativist movements, the new Christian right is supported by only a small fraction of the bourgeoisie.

In the absence of detailed research on this question, my impression is that capital divides by geography and constitution. The capitalists most supportive of the NCR (and of earlier right-wing organizations such as the John Birch Society) are located in the sunbelt: the Hunts of Texas, the Coors of Colorado, Ted Collier of Alabama, and J. H. Pew, the founder of the Sun Oil Company. This distribution represents something of a reversal of an earlier pattern. In the last century as now, the Republican party was divided into progressive and conservative factions. The progressives were to be found in business in the Mid-West and the sunbelt:

> the progressivism of these relatively well-to-do men, as well as that of many of their lower-middle-class followers on farms or in business, consisted in seeking to restrain the large corporations which were stifling individual initiative. Their ideal society was the vanishing America in which the native-born rural or urban entrepreneur set the tone for the whole society, and they sought to maintain that society through government action to inhibit the growth of large corporations and trusts. (Lipset 1969: 315)

The major changes in the economy since the turn of the century have reversed relative ideological positions. Where old north-eastern capital was once fervently opposed to state intervention, it has now learnt to live with government regulation because it benefits from government contracts. Old capital was big enough to be able to accept labour unions and the improved wages and conditions they demanded. It has become bureaucratized and much of its administration has passed into the hands of professional managers. Large corporations are now controlled by 'college-educated men and the scions of established wealth, rather than by the relatively less educated founders of new firms' (Lipset 1969: 316). The sorts of business men who were once on the progressive wing of the Republican party are now its most aggressive conservatives, campaigners for deregulation and free market forces. The old capital of the North East is conspicuously unsupportive of the new right and even less keen on the NCR. No family better represents old wealth than the Rockefellers and far from supporting religiously particularistic movements, John D. Rockefeller gave a large gift to the newly formed World Council of Churches—the fundamentalists' 'great satan'—in the 1950s (Till 1971: 235).

Nelson Rockefeller, who was Governor of New York and Vice-president to Gerald Ford, is frequently envoked by NCR activists as the exemplar of all they detest in rich, liberal, cosmopolites.

As the person most heavily involved in early fund-raising for both secular and Christian elements of the new right, Richard Viguerie is presumably in a good position to judge when he follows his listing of supporters of the new right by saying:

> You may notice a missing element here: big business. I'd like to say something here about the so-called support for conservative causes by big business. As far as the New Right is concerned, it never existed. It is not true that what is good for General Motors is good for the country—or that what is good for big business is automatically good for conservatives. The New Right rejects both propositions. This is one of the things that distinguishes us from the Republican establishment (1981: 107).

In 1982 conservative Christians spent about $3,000 dollars taking over the Fairfax County Executive Committee of the Republican party. Four years later liberal Republicans spent $53,000 regaining control of the local party. The money came from land developers. In county after county the representatives of capital have supported the 'moderate' party establishment against the religious right. In the explanation of one Christian Voice activist:

> The majority of the big money men are three Martini Episcopalians who belong to the Country Club and drive a Rolls or a Jag or something and they despise these unwashed low-income Christians coming in singing their hymns and trying to take the Party away. You have to understand that these guys belong to the First Church of the Bottom Line! That's what they care about.[2]

While the representatives of capital might well see long-term advantages from a major shift to the right, they see more immediate, predictable, and calculable advantages from doing business with the devils they know. Shortly after the November 1986 elections, a number of business political action committees made donations to the liberal Democratic senators who had defeated the same conservative Republicans those committees had supported in the election! Archer-Daniels-Midland is America's largest agricultural processing concern. Since 1979, it has given more than half a million dollars to various legislators. But in

[2] All quotations which are not accompanied by a published source are from transcriptions of interviews conducted in spring 1983, 1986, or 1987.

addition to supporting such well-known conservatives as Jesse Helms, Gordon Humphrey, and Phil Gramm, it has also made sizeable donations to liberals such as former House Speaker Tip O'Neill, Senator Tom Harkin and Representative Thomas Downey. 'Even while ADM and Andreas [the major shareholder] were bankrolling Percy in his Senate re-election bid, the Andreases gave his opponent, Democrat Paul Simon, $5,000 with an additional $1,000 kicked in by the ADMN PAC to boot' (Fumento 1987: 36).

It is worth adding that even for those capitalists most active in supporting the new religious right, politics takes second place to evangelism. To quote one NCR fund-raiser:

> Take Bunker Hunt. He's given an enormous sum to political causes compared to the rest of our funders. This man over the years has given about 20 to 30 million dollars to Bill Bright [of Campus Crusade for Christ] and he's given maybe 10 per cent of that amount to political causes. Most of these guys give their money to ministries. Ted Collier, a fat cat from Alabama, has given over a million bucks to Falwell's Old Time Gospel Hour but maybe he'll pop for five or ten grand to Moral Majority.

As an additional caution to accepting too readily either a divide-and-rule or an opium-of-the-masses account of the use of religious issues, it is worth noting that even when we find conservative religious activists advertising the secular consequences of their enterprises, we cannot simply assume cynical manipulation. Clearly the value of being able to recruit non-religious conservatives to support a socio-moral crusade, by pointing to consequences which might suit them, is an important rhetorical device in any socio-moral crusade. When Wilberforce and other British evangelicals were fund-raising for the establishment of mass schooling, they presented different justifications to different audiences. Convinced evangelicals were told that mass literacy would make it more likely that the lower classes would be converted to the saving faith. Secular conservatives were told that teaching the masses to read would allow them to be fed improving literature which would deflect them from imitating the French and revolting (McLeish 1969). Rather than being evidence of ruling-class cunning, this seems more like opportunism on the part of true believers who use whatever means are available to gain support for their goals.

So one has activists with different interests and one has activists who are willing to recruit support as widely as possible for their campaign by advertising different consequences to different potential supporters. In so far as non-religious conservatives are prepared to use religion as a means of diverting political choices from economic issues, most will seek to define religion in the most inclusive, i.e. universalistic, terms so that the advantages which accrue from such a mobilization are not offset by the disadvantages of social instability and sectarian conflict. Here the interests of the secular conservatives and the more perceptive and pragmatic promoters of the NCR coincide.

The point of the above is simple. There is no doubt that the promotion of the NCR served conservative interests. One can go further, as I will later, and suggest that the NCR was of *more* use to the mainstream of the Republican party than it was to its own ideologically committed supporters. However, there is no need for that recognition to take the form of manipulationist theories (of Marxist or non-Marxist bents). One need only acknowledge that involvement in any large-scale social mobilization will spring from a variety of motives and will be accompanied by a variety of rhetorics. In so far as the bourgeoisie can be identified as having a Lipsetian interest in the use of religion, it should be noted that the costs—in terms of increased tension—of reintroducing religious particularism (especially in a democracy) will probably outweigh any advantages. Only those members of the bourgeoisie who are personally committed to some religious particularism will consider the likely results worth while. And, both for such people and for more cosmopolitan sections of the ruling class, every attempt will be made to make the non-economic values around which the mobilization is to be based as broad and as inclusive as possible.

THE HOLY TRINITY

We should be cautious of following journalists into the trap of telling the story of the new Christian right in terms of leading personalities. None the less, the catalyst for the NCR and the name of the leading organization—Moral Majority—were both supplied by a triumvirate of professional right-wing organizers: Richard Viguerie, Howard Phillips, and Paul Weyrich.[3] With financial

[3] While Viguerie may have been the pioneer of direct mail, his political judgement is usually bad and he charges a lot for his services. For these reasons, many NCR activists in private conversation are now quite disparaging of him.

support from Joseph Coors, the head of one of America's largest breweries, Weyrich had already created a number of right-wing political action committees (hereafter PACs). Although PACs have a long history in American lobbying, their number has multiplied considerably since post-Watergate legislation limited the amount of money which any individual could give to a particular political candidate (*Congressional Quarterly* 1982). Both Phillips and Viguerie had worked for Young Americans for Freedom, an organization formed in the early 1960s by William F. Buckley.

Between the three of them, Viguerie, Weyrich, and Phillips created a mass of organizations so interlocked that 'any diagram . . . looks like an octopus shaking hands with itself' (Davis 1980: 21).[4] What distinguished the triumvirate from other right-wingers was their frustration with the Republican party and their desire to build a new conservative majority of people who were opposed to the social and moral liberalism which had pervaded America in the 1960s. They were, of course, right wing on the more traditional issues of defence spending, foreign policy, and the reduction of government interference in the economy. But at times, as Viguerie's comments on big business suggest, their politics seemed almost to have more in common with the old Democratic party populism of William Jennings Bryan than with the conservatism of Barry Goldwater, whose 1964 presidential campaign has been claimed as an inspiration by many new rightists. They certainly had little in common with the liberal and cosmopolitan Republicanism of Nelson Rockefeller.

Although Viguerie and Weyrich often describe their activities as bipartisan, it is more accurate to say that they were alienated from

[4] Although it breaks the chronology, the extent to which new rightist organizations overlap is shown by the membership of the Council for National Policy, created by Tim La Haye in 1980. La Haye is an independent fundamentalist Baptist minister who had built a large and successful ministry before becoming active in Falwell's Moral Majority Inc. In addition to Phillips, Weyrich, and Viguerie, members included Bob Billings and Ed McAteer of Religious Roundtable; conservative patrons Holly and Joseph Coors, and Herbert and Nelson Bunker Hunt; television preachers Pat Robertson and James Robison; Terry Dolan of National Conservative PAC; Reed Larson of Right to Work; Connie Marschner of National Pro-Family Coalition; Congressman Larry P. MacDonald of the John Birch Society; Phyllis Schlafly of the Eagle Forum and STOP-ERA; Jerry Falwell and Richard Dingman of Moral Majority Inc.; and a number of less well known conservative activists. Details of conservative political organizations over the last 26 years can be found in the excellent monthly *Group Research Report*, published by Wes McCune, 527 Woodward Building, 15th St NW, Washington DC 20005.

the Republican party; their politics certainly had little chance of support in any part of the Democratic party outside the southern 'Dixiecrats'. Their initial lack of commitment to the Republicans showed itself most dramatically in their willingness to support third-party conservatives. In 1974 Viguerie and Phillips formed the Campaign for the Removal of the President, motivated by dislike, not for Nixon's Watergate evasions, but for his liberalism. They were even less happy with Gerald Ford, especially when he appointed Nelson Rockefeller as his Vice-president. In 1973, Viguerie accepted the job of raising money to settle the 1972 campaign debts of George Wallace, the Democratic ex-Governor of Alabama. Two years later Viguerie and Weyrich talked with Wallace about another presidential campaign but the injuries he had sustained in the attempt on his life in 1972 dissuaded him from accepting their offer. In 1976 Ronald Reagan alienated the triumvirate by choosing the liberal Richard Schweiker as his running mate and, in disgust, Viguerie offered his services to the American Independent Party. In the run-up to the 1980 elections, they again cast around for a genuine conservative, trying to enlist ex-Democrat John B. Connally, before having to accept the inevitable and support Ronald Reagan. As an aside—the point will be pursued later—the choice of Connally was quite deliberate. During his presidency Nixon had deliberately cultivated conservative southern Democrats in the hope of building a new majority party of conservatives. Connally was groomed by Nixon as his successor but the Watergate débâcle put a temporary halt to what was intended to be a major realignment in American politics.

In his characteristically folksy style, Viguerie argued that:

> Conservatives should work for the day when the November contest is between a conservative Democrat and a conservative Republican. Then we can go fishing or play golf on election day knowing that it doesn't matter if a Republican or a Democrat wins—it's only important that a conservative wins. (1981: 89)

Apart from their 'bipartisanship', Viguerie, Phillips, and Weyrich also differed from previous conservatives in trying to construct a political movement by creatively combining an array of single-issue constituencies into multi-issue movements. The interest in creating a new conservative majority based on single-issue campaigns coincided with a new public awareness of the strength of conservative Protestantism. It also coincided with an increase in the sense among conservative Protestants that the only way to preserve their world was to fight back. It was thus natural that Viguerie, Phillips, and Weyrich should become interested in conservative Protestant leaders. In the words of Viguerie's *Conservative Digest*,[5] they:

decided that the millions of fundamentalists in America were a political army waiting to be mobilized. The two leading groups, the Moral Majority, headed by the Rev. Jerry Falwell, and the Religious Roundtable, led by Ed McAteer, came out of meetings attended in early 1979 by Weyrich, Phillips, McAteer, Falwell and others. According to McAteer, who introduced Phillips to Falwell, the term 'moral majority' was coined by Phillips and first used publicly by Weyrich in a presentation to Falwell and his associates. (in Phillips 1982: 190)

THE MECHANICS OF MOBILIZATION

We can see why conservatives should have wanted to mobilize America's fundamentalists. The account presented in Chapter 2 of the increasing secularity of cosmopolitan America, and the increasingly frequent interventions of the state in affairs dear to fundamentalist hearts explains why a number of fundamentalist leaders were interested in becoming more politically active. The ways in which a viable movement was created will now be examined.

A vital element of the mobilization was the technology of direct mailing. Viguerie tells the following story of his own awakening to the possibilities of mail solicitation. While working for Young Americans for Freedom, he was supposed to ask rich people for money. He found this rather embarrassing, took to writing begging letters, and found his vocation. His first list of names and addresses was copied down from the Congressional record of all those who had given donations to Barry Goldwater's 1964 campaign.[6]

Modern computers can store and print vast numbers of names and addresses. They can also print 'personalized' letters. A basic

[5] *Conservative Digest* is a glossy pocket-sized monthly modelled on the *Reader's Digest*. Founded by Viguerie, it was sold to conservative activist William Kennedy sometime around 1985.

[6] Although Crawford (1980: 42–77) denies the originality of Viguerie's direct-mail techniques, he acknowledges his lead in what is now a crowded field and gives an excellent description of some of the techniques.

text can be altered in detail to suit different recipients. A realistic facsimile of a personal signature can be attached. The most recent technical development allows a whole letter to be printed as a facsimile of handwriting.

The skillfully written letter is long enough—three pages or more—to persuade the recipient that he is an important citizen. It is computer-personalized, often with his name dropped in several times. Sometimes it includes what looks like hand-scrawled postscripts, as if the senator or other busy figure sending it took time out from his full schedule to make this special point to this individual alone. It conveys a spirit of conspiratorial comradeship: sometimes it bears a 'confidential' stamp, as on a secret shared only among intimate friends (Furguson 1986: 144–5).

Direct mail is used in the following way. Suppose that one wants to campaign against the signing of the Panama Canal treaty. The machine is fed a basic text which explains why the treaty is a bad thing. The text will also ask for a donation to cover the costs of the campaign. Part of the letter may consist of three ready printed, stamped, and addressed postcards which warn the recipient's congressman and senators not to vote for the treaty. This package is posted to all those people on the records who are likely to be opposed to the treaty. If enough recipients tear off and mail the cards, members of Congress are flooded with expressions of opinion hostile to the treaty.

One important advantage of the technology is that its records can be updated cheaply so that the initial list is subdivided to show those who sent in donations of various sizes. The lists can also be cross-referenced so that the computer can generate the names and addresses of those who are anti-gun control, or anti-Canal treaty, or anti-ERA (as the equal rights for women constitutional amendment is popularly known), or anti-school busing.

Although such technology is not likely to change radically the opinion of the recipient, it is good for alerting people to salient issues which might otherwise have escaped their notice. It also allows the creation of abstract political constituencies. Although the degree to which any person is involved with the Democratic party or a trade union, for example, clearly varies considerably, the assumption of traditional politics is that the mobilizing organization has to appeal to a coherent package of beliefs and values. The 'constituency' mobilized by direct mail is quite different. Although overlap is expected (and empirically is found) there is no need to discover or mobilize people who are anti-gun control *and* anti-Canal treaty *and* anti-school busing. Each of these issues can be treated discretely. Making donations to a PAC which is active on one issue and sending in pre-printed cards expressing an opinion on that, does not commit the actor to endorsing a complete philosophy. It is not clear from the writings of Viguerie and other NCR strategists whether they appreciated the value of this approach, but subsequent studies suggest that— irrespective of the degree to which a single-issues approach was likely to succeed—the alternative would have failed. Shupe and Stacey (1982) and Himmelstein and McRae (1984) find significant differences on issues within what would appear to be the natural NCR constituency. In particular, many people who shared the NCR leadership's attitude to abortion were nowhere near so unified in their attitudes to the equal rights for women constitutional amendment (ERA).

Direct mail has the related advantage that it can create the appearance of an active constituency without demanding that any one actor make much of a commitment. In many ways, it is ideally suited to a privatized world. People do not have to leave their homes, join anything, or give up much of their spare time (although some people will be sufficiently galvanized to do those things). Just as direct mailing can raise large sums from a large number of small donations, so it can mobilize a powerful constituency from very small acts of commitment from large numbers of recipients. Although I would not want to endorse the image of the modern world as a 'mass society' composed of isolated and anomic individuals, it seems unquestionable that the erosion of residential and occupational communities has created a problem for political movements which depend on more traditional modes of mobilization. Although the sort of involvement that direct mailing can engender is both qualitatively and quantitatively different from that stimulated by more traditional methods, it is none the less a valuable technique for the modern world.

Having said all that, it is worth adding the rider that direct mail is not the alchemical formula that many politicians and journalists believe. It is expensive to operate and the first few mailings on any list will cost considerably more than they raise. Only once a new list has been pared down will it be profitable, and even then its useful life is short. People move, die, or develop new interests. With some justification, Tina Rose (1983) argues that the main winners in direct mail are the direct mail companies themselves. But whatever its value as a method of fund-raising, direct mail is a good method of consciousness-raising for those people sympathetic enough to read it.

TELEVANGELISM AND CONGREGATIONAL NETWORKS

The triumvirate's choice of Falwell was not accidental. Jerry Falwell was an independent Baptist fundamentalist from Lynchburg, Virginia, who had created, in addition to his own congregation, an independent school, a Bible college (which is now a liberal arts university), and a popular television gospel show. Although Falwell's percentage of the total audience for such shows placed him only sixth, with about half the audience which watches Oral Roberts or Rex Humbard (Hadden and Swann 1981: 51), his show none the less gave him a ready-made audience for any attempt to politicize fundamentalists. Although other televangelists, as Hadden and Swann call them, were less forthright about promoting a particular organization, they too took the opportunity to express their concerns about social, political, moral, and legislative change and helped to promote the aims of the NCR. Marion G. 'Pat' Robertson, head of the Christian Broadcasting Network and host of the *700 Club* could spread his conservative religio-political views to huge audiences through his shows and his periodical *Pat Robertson's Perspective*, a monthly with 247,000 subscribers (Conway and Siegelman 1982: 62).

As Liebman (1983a: 52–73) has correctly pointed out, much attention has been given to the role of direct mail and other forms of parapersonal communication such as televangelism, and the equally important congregational networks have been neglected. Posing the question in terms of trying to explain why the Moral Majority Inc.[7] was more successful than competing NCR organizations in creating a viable organization, Liebman points to the homogeneity of background of the chairmen of the state chapters of

[7] On 3 Jan. 1986 the name of the main organization was changed from Moral Majority Inc. to Liberty Federation. However, some sub-units still use the Moral Majority label and many activists (including the telephonist!) still use that name. When speaking of Falwell's movement generally and specifically about it pre–1986, I will use the old name.

the Moral Majority. All the forty-five chairmen whose denomination could be ascertained were Baptist ministers. Twenty-eight of them were fundamentalists affiliated to the Baptist Bible Fellowship. Those who took the lead in the Moral Majority—Jerry Falwell, Tim La Haye, and Greg Dixon—also had in common their success in having founded and built up 'super churches'. Liebman makes two points about this shared background, in addition to the obvious one that it guaranteed they shared the same religious beliefs. The first is that independent Baptist fundamentalist clergy are, in the jargon of American management theory, 'self-starters'.[8] The Fellowship in which they are linked does not guarantee them a living or a career.

> The fates of their churches depend largely on their success in local evangelism. For fundamentalist pastors, aggressive soul-winning is more than a matter of conviction; it is an organizational imperative. Their work in recruiting congregants through telephone chains and door-to-door visits sharpened the skills of grassroots mobilization. (Liebman 1983a: 68)

They also enjoy considerable pastoral autonomy and hence, provided they can carry their congregations with them, may pursue lines of action without being censured or controlled by a denominational organization.

The shared background in independent ministries explains the entrepreneurial attitudes and managerial skills of the Moral Majority state chairmen. It also explains the ties of fellowship which bind them together. Although in no sense subordinate to an overarching organization, independent Baptist pastors, especially in the first stages of church-planting, do depend on assistance from other pastors and congregations. Falwell, for example, makes a practice of preaching at churches founded by graduates of his Liberty University and Seminary. By sharing 'fellowship' and assisting each other, independent Baptist pastors develop a strong network of interpersonal ties.

Finally, it should be noted that the involvement of fundamentalists in politics did not begin with the inauguration of the Moral Majority Inc. Greg Dixon, a leading Moral Majoritarian, had long been involved in local disputes in Indiana and many of the other pastors who were appointed as state chairmen had considerable

[8] Bruce (1986b) makes similar observations about the independent ministries of many British militant Protestant leaders.

experience with city and state campaigns on behalf of fundamentalist schools and anti-abortion crusades. In one sense, the formation of the Moral Majority Inc. or Religious Roundtable was merely the linking together at the federal level of a series of small movements and organizations which had been campaigning on particular elements of what became the NCR platform. In this sense, the 'newness' of the new Christian right lay in new organizational forms and a change of gear.

The general point is this: while direct mail and televangelism provided network links between leaders and followers, the already existing ties between a large number of Baptist fundamentalist pastors provided a ready-made framework for the mobilization of a leadership cadre.

NETWORKS IN SOCIAL MOVEMENT THEORY

It is often the case that a research tradition develops in a series of pendulum swings; in reacting to the weaknesses of one generation, the next over-compensates, and it takes a third to restore a balance. Social movement research is no exception. By the 1970s, a number of scholars were arguing that previous research had overlooked a series of important problems. The theorists discussed in the opening section of the first chapter were so concerned to find the structural causes of collective action that they assumed that people in a common position of objective 'strain' simply 'reacted'. Strain was the common stimulus; collective action was the shared response. Little thought was given to the problem that the number of people in the common objective position was always far greater than the number who engaged in the supposed response. As sociologists rediscovered an interest in how people perceived the situations in which they found themselves, attention shifted to the ways in which collective definitions of situations were developed, and to the social processes which might explain why some people rather than others were mobilized by such definitions. Analysts began to take an interest in social networks.

A number of studies of recruitment to social and religious movements argued that a major variable in such recruitment was the existence of social networks (Lofland 1985; Snow *et al.* 1980). It seems clear that many new ideas spread most effectively when they are transmitted along lines of already existing friendship and kinship links. Where such ties are absent, they are often created. Hence the considerable efforts which trained recruiters such as the Mormons and Witnesses put into making friends with potential converts. There is obviously considerable merit in this proposition. However, some studies seem to have regarded bold theoretical innovation as more important than sensible analysis and have proposed a new structuralism as reductionist and unsatisfactory as the old structuralism of Parsons and Smelser. Snow *et al.*'s 'micro-structuralist approach to differential recruitment' suggests that whether or not an individual will participate in a movement 'is largely contingent on the extent to which extra-movement networks function as countervailing influences' (1980: 792–3). Recruitment is 'a function of how tightly individuals are tied to alternative networks and thus have commitments that hinder the recruitment efforts of social movement organizations' (1980: 794). Wives, children, jobs, and mortgages are the sorts of commitments which Snow *et al.* have in mind.

Snow *et al.* add a structuralist twist to network analysis. The question of how receptive people are to some ideological innovation is divorced from the nature of the innovation and reduced to a question of availability as 'receptivity' is defined simply as position in a network. In their account of the importance of social networks, Stark and Bainbridge add an element of deprivation theory (1985: 314–24). They argue that social networks are central because the development of affective bonds (that is, 'friendship'!) with movement members is one of the direct rewards of participation. Some joiners are deprived of social ties. A movement offers a solution to that deprivation.

Taken generally there is much merit in both these suggestions but when the additional insights are given too much explanatory weight, they become reductionist in that they involve unwarranted disregard of actors' motives. People who join new religions or social movements claim they do so because they have come to believe in the ideology or the cause. That is, their explanations focus on cognitive concerns. The sociologists substitute an alternative. In the Snow *et al.* case the reason for joining is an absence of countervailing influences (construed rather mechanically as observable social ties to non-movement members). In the Stark and Bainbridge case, it is desire for friends.

Networks are important for two reasons: knowledge and belief.

The first stage in any mobilization is the spread of information. People cannot convert to something of which they are unaware nor can they join movement organizations of which they have not heard. Where social networks often have the advantage over less personal means of communication is in *vouchsafing* the information they transmit. Pre-existing personal relationships of trust confer added authority on the messages they transmit. That I have trusted my friend on previous occasions is a good reason to take seriously and consider sympathetically what he tells me now. Put like this, it should be clear that networks do not operate independently of cognitive factors. Rather they have the importance they do because so much of what we know about the world has to be taken on trust.

Putting it like this has the added advantage that it does not radically separate personal networks from impersonal ones. Personal trust is not the only source of authority (nor, incidentally, is it an invariant property of network ties). Institutions and organizations may also possess authority and impersonal channels of communication will vary in the degree to which their transmission of some message or innovation increases the chances of adoption. Because I have found the *Guardian* newspaper to be generally and consistently supportive of the causes and positions I support, I find it plausible, and, for some innovations, more plausible than my friends.

As some of the excellent and sadly neglected research in the diffusion of innovation tradition has shown, the vouchsafing element will vary in salience depending on the nature of the new information or ideas. For certain sorts of innovation, personal networks are little more effective than impersonal ones because the innovation can be carefully tested before full adoption. The spread of hybrid seed corn offers a good example. Most farmers adopted the new corn tentatively, planting a small amount in the first year, and comparing the results with their established seed. Only as they became convinced of improved yields did they increase their reliance on the innovation (Rogers 1962: 84–6).

It would be a mistake to suppose that ideological innovations are of an entirely different nature. It seems clear from sensitive studies of recruitment to new religious movements, for example, that people do adopt these innovations gradually and with a certain degree of experimental scepticism. Only when they have an increasing sense that the new beliefs, values, and practices are 'working' for them, do they become 'true believers', or, in the diffusion of innovation language, move from evaluation to full adoption (Bromley and Shupe 1979). None the less, it seems clear that religious or political innovations are far less readily testable, or, to be more sociologically precise, the testing—the sense of knowing whether or not an ideological innovation is working—depends far more on accepting a particular social construction, and hence depends far more on authority and trust than does the evaluation of a technical innovation.

Hadden and Swann (1981: 13–14) have made an important point about the nature of communication between televangelists and their audience which relates to the above observations on the importance of personal affective bonds in establishing trust. Although televangelists use media that are 'mass' and hence impersonal, they have skilfully used available technology to produce elements of, and a considerable appearance of, personal communication. Many shows invite the audience to write or phone in with requests for prayer and with stories of prayer answered. Prayer requests are read over the air. The computers which write to members of the audience are sufficiently sophisticated to be able to inject into standard texts personal elements such as the name of the recipient, some reference to the problem which they mentioned in their letter, and the promise of prayer for that problem. Although television is a mass medium, close-up camera shots and good reception can make the 'stranger in the box' in the corner of the living room seem closer to the viewer than their actual pastor at the far end of a church building. To use Hadden's phrase, state-of-the-art televangelism is turning an impersonal medium of communication into a 'parapersonal' one.

Pre-existing networks were very important to the rise of the new Christian right. The two key messages—'America is degenerate' and 'we can and should act to reverse that degeneration'—were transmitted to the existing audiences for evangelistic television and radio shows, and through the fraternal networks which linked fundamentalist pastors to each other and to their congregations. It would be a mistake to exaggerate the novelty of the innovation that was being promoted. Christians were not becoming Moonies; liberals were not becoming conservatives. The vast majority of people in these networks were already conservative evangelical

Protestants. The message was directed precisely at these people. The innovation was to ask that they become politically active. Equally well, the novelty of this request should not be played down (as it is by resource mobilization theorists such as Miller). Given the long tradition of fundamentalist Baptist pietistic retreat from the world, registering to vote, subscribing to political funds, campaigning for candidates; these things were novel. Hence the same problem which dogs any innovation was present. Those who heard the messages had to ask themselves and each other 'Why should we believe this person?' That the person in question had already acquired trustworthiness and authority in the eyes of the audience was important to the acceptance of the innovation. Thus we can conclude that the decision to try to mobilize socio-moral conservatives through the audiences for televangelism shows and through church networks was a sensible one.

CONCENTRATION AND MOBILIZATION

The explanation of the rise of the NCR involves an account of those forces which caused people to be potentially open to recruitment to such a movement, and of how they were recruited. The first point can be summarized by arguing that the late 1960s and 1970s saw an increasing penetration of peripheral regions and subcultures by cosmopolitan culture, and an increasing liberalism in the values and practices of the 'centre'. As Kevin Phillips puts it, the world of Manhattan, Harvard, and Beverly Hills was being exported to Calhoun County, Alabama, and Calhoun County did not like it (1982: 187). There have also been a related series of socio-economic changes which have increased the power and profile of the South (and the sunbelt generally) so that the dog-in-the-manger attitude which has prevailed since the Civil War and Reconstruction has turned into a sense that southerners have a significant part to play in the politics of the nation.

The second part of the explanation—the resource mobilization story—concerns the involvement of a number of skilled political activists, their ability to see a constituency, and their knowledge of the techniques that could be used to mobilize it. Direct mail, televangelism, and ministerial associations were all used as channels along which a particular interpretation of America's recent past and present could be transmitted.

The third element of the explanation concerns the feedback loop which ties success to motivation, so that the sense that there was a good chance of making some impact became a crucial determinant of the nature of the campaign and the enthusiasm which it attracted. I will argue that there are distinctive features of the American political and administrative structure which make movements such as the NCR much more likely to succeed in America than in, for example, Britain.

To an extent the comparison is artificial. There are so many differences in the history of religion and politics in the two settings that it is difficult to imagine Britain producing a large market for a social movement mobilizing around socio-moral issues. Britain is far less religious than America. About 60 per cent of Americans claim church membership and regularly attend church. Only 17 per cent of people in Britain claim membership and the most recent figures for church attendance suggest that only some 11 per cent of the population are church-going.[9] None the less, there is heuristic value in imagining in Britain a large constituency of conservative Christians who have transcended their tradition of Protestant–Catholic conflict and are prepared to campaign on shared socio-moral positions. Comparative observations on the structures of British and American polity will be presented to show why, even if there were a British market for a 'new Christian right', it would not follow the American model. In this way, the structural features of the American context, summarized in Table 1, which were

TABLE 1 *Contrast between British and American political systems*

	Britain	America
Government	National Centralized	Federal Diffuse
Offices	Appointed	Elected
Mass media	Closed Centralized	Open Diffuse
Parties	Strong	Weak

[9] The best recent statistics for British church membership and attendance have been produced by Peter Brierley and associates, MARC Europe, 6 Homesdale Rd, Bromley, Kent BR2 9EX.

especially conducive to the rise of the NCR will be highlighted. In such a brief summary there will, of course, be an element of caricature and exaggeration, but what follows nevertheless represents a picture of contrasts generally accepted by most political scientists (Vile 1976; Denenberg 1984; Finer 1982).

Government

British government is national and centralized. Decisions made by the Westminster Parliament are paramount. The Stormont Parliament, which gave some autonomy in domestic matters to Northern Ireland, was closed down in 1972. The Scottish Grand Committee of the Westminster Parliament has so little control over Scottish affairs that many Scottish members do not bother to attend its meetings. Although there are various tiers of local government, these are subordinate to Westminster; so much so that the national government is able to alter the structure of local government to remove troublesomely independent Labour-controlled authorities, as the Thatcher Government did with the Greater London Council. The American system has certainly moved in the direction of more centralized decision-making, but state legislatures still retain considerable power. Each state has its own executive, judiciary, and bureaucracy, and the most important pieces of legislation—constitutional amendments, for example—have to be ratified by the separate states.

The centralized nature of British politics means that metropolitan culture is forced upon the peripheries. Despite considerable local opposition, law on homosexuality in Northern Ireland was brought into line with the more liberal law of the rest of the United Kingdom. Sunday opening of public houses was also imposed by an 'Order in Council'. Whatever decision the Government makes about Sunday trading will be made in London and will pay scant regard to the interests of Scottish or Welsh sabbatarians.

Given that schools play an important part in helping or hindering parents in the transmission of their values to their children, educational policy is of considerable concern to the NCR. In Britain, educational policy is made in London and imposed on the rest of the country (although Scotland retains some autonomy). Until 1986, decisions about funding were divided between central and local government, with Westminster providing half the funds. However, the Thatcher Government's dissatisfaction with protracted pay negotiations with the teachers led it to take complete control of salary negotiations. Although education in Northern Ireland is administered by its own civil service department, the contents of syllabuses have long been determined by national Inspectors and Examination Boards. In America, education is a state and county matter. This degree of local control means that cultural groups which are strong in particular areas can do more than their British counterparts to ensure that their values are respected.

Federalism not only has the consequence that it permits the survival of regional cultures; it also encourages the bearers of such cultures to act in elections. Baptist fundamentalists in North Carolina cannot pack the federal congress with fundamentalists but they can hope to put conservative Protestants into the state legislature.

Finally, the strong cabinet model of Westminster politics amplifies the power and cohesion of the centre while the American separation and balance of powers between presidency, Congress, and judiciary weakens the centre and permits minority and regional groups to hope to retain or gain some areas of freedom from central control.

Public Administration

British public administration is élitist and paternalistic. A large number of important offices—seats on the judiciary, for example—are filled by appointment. Senior positions in the bureaucracy, central and local, are filled by professionals who do not normally rotate with changes in the party in power. Complex and contentious issues are weighed by Royal Commissions composed of the 'great and the good'. In contrast, Americans like to elect:

> It is not uncommon for the electors to face a 'long ballot' containing the names of from fifty to one hundred candidates—to elect half a dozen state officers apart from the governor, to elect state comissioners and judges, state treasurer and state attorney, as well as their mayor, their councillors, the members of their local school board, the city court judges, their tax collectors, and many more. (Finer 1982: 227–8)

Ironically, there is an inverse relationship between the number of elections and the interest taken in them. The American turn-out is lower than that of all but one of the Western European countries. In the 1978 mid-term elections, only 35 per cent of the population went to the polls (Neely 1981: 26). In 1986, just over 37 per cent of eligible adults voted. In such circumstances, there is obvious opportunity, and hence encouragement, for any well-organized group to become an electoral force.

To date, the élitist and paternalistic nature of British public administration has been seen as a problem only by the left, quite reasonably, given the shared upper-class social background of those chosen to fill important positions. However it is worth noting that, while judges, for example, generally support positions which are moderately conservative, groups which wish to pursue particular socio-moral issues associated with the right are as impotent as the left when it comes to bringing direct pressure to bear. Judges will be appointed (from the ranks of those qualified, and qualification is more narrowly defined in Britain than in America) by the Lord Chancellor, who is a senior government politician with considerable legal experience. Given that the character of the Lord Chancellor and the stand of the two main parties on socio-moral issues are not normally major issues in general elections, there is little chance of any particular interest group being able to use electoral power to change the composition of the judiciary. In November 1986 Rose Bird, the Chief Justice of the California Supreme Court, was voted off the bench after a long campaign of criticism of her opposition to capital punishment.

Mass Media

The openness of American mass media has already been mentioned, and the contrast with the British system of a small number of channels need not be laboured. For the purposes of this discussion, a crucial difference is the ability of American interest groups to buy either time or whole channels to air their views. Equal access legislation has failed to prevent money becoming the main determinant of access to the American air waves. In Britain the publicly financed British Broadcasting Corporation (BBC) and the Independent Broadcasting Authority (IBA) (which acts as a watchdog over the independent television companies) prevent ideological advertising (the BBC does not carry any paid advertising), and operate a fairly strict notion of balance which promotes a moderately conservative centrist view of most issues. They also allocate electioneering time to the main parties, rather than to individual candidates. Hence there is no possibility in Britain of any particular pressure group buying air time to attack or support any particular candidate.

Political Parties

The final important structural feature which distinguishes Britain from America is party cohesion. At first sight both countries are similar in having 'first past the post and winner takes all' elections, two main parties, and a history of the failure of third parties to have enduring impact. However, the superficial similarities disguise vast differences. In Britain, candidates for elections are party candidates, chosen and endorsed by the party. Where the central party organization disapproves of a candidate selected by the local branch, it can impose its own candidate, as the Labour party did in the Knowsley North by-election in 1985 and in Nottingham before the general election of June 1987. It can also expel party members, even those who hold high office in local government; the expulsion from the Labour party of the Militant Tendency supporters on the Liverpool Council is such an example. The Conservative party exercises even greater control. Local party branches choose their candidate, but they must select from a list of those who have been examined and approved by central party headquarters.

The results of British elections rarely depend on the personal or ideological qualities of local candidates. The success or failure of most candidates is a function of the national standing of the parties to which they belong. Election policy, advertising, and funding are all matters for the national party, and legislation which firmly limits the amount a candidate may spend maintains this oligarchic control.

There is an element of the illusory about the two main American parties. Candidates are not chosen by the parties—they need not even be members of the parties they claim to represent—but by the electorate in a primary election.

> Consequently, the decision as to who is—and hence what is—a Republican . . . has been taken from the hands of the party bosses, whether at local or at state level or at federal level, and vested in local populations: and this means that Democratic or Republican means just what this or that locality, in this or that year or set of circumstances, has decided they shall mean. And that varies all over the country from election to election. (Finer 1982: 228–9)

Only a small proportion of the (since the advent of the television campaign, considerable) funds required to fight a contested election comes from the national party organization. Most is raised by the candidates themselves. As a result the victor owes little or nothing to the party machine.

The lack of party discipline is clearly visible in the relative proportion of 'party votes': those divisions in the legislature when 90 per cent of one party votes against 90 per cent of the other. Almost all Westminster votes are party votes. Only very rarely do party whips allow their members a free vote and, because most members owe their seats to party rather than personal popularity, and similarly acquire promotion through loyalty rather than through seniority, very few Westminster members break ranks and vote against their own side. In contrast, only 5 per cent of congressional roll call votes between 1960 and 1970 were party votes (Turner 1970: 239). The point is dramatically made with the voting record of the late Democratic Congressman Larry P. MacDonald of Georgia, a leader of the NCR who voted against Jimmy Carter, Democratic President and ex-Governor of Georgia, more often than any other member of the 95th Congress.

The lack of party discipline in American politics is a major incentive for pressure groups and social movements to become electorally active. In Britain there is little point in trying to replace a pro-abortion candidate of any party with an anti-abortion candidate. Even if a socio-moral interest group could pack a local constituency branch and have its own candidate nominated (which itself is very unlikely), the national party could step in and replace any candidate who threatens to campaign on issues which are not part of the national party's manifesto. And even if such a candidate could be returned, the impact on Westminster would be minimal. The legislative agenda and voting in Parliament are firmly controlled by the national party officers. As Grant puts it, members of the British Parliament do not legislate, they *legitimize*: 'Ordinary MPs are under great pressure to vote as their party leaders instruct . . . the significance of constituency interests, pressure group influence or personal involvement is considerably diminished' (1979: 64).

The American primary system of candidate selection offers opportunity for choosing an individual with the correct socio-moral positions. As will be seen below in the discussion of the activities of the NCR, access to the media permits pressure groups to introduce their issues into a campaign. The congressional system allows individual legislators to initiate legislation.

In the long run, even the American system tends to filter out particularisms. Socio-moral legislation may be introduced but not attract sufficient support to survive the convolutions of a system well designed for stasis. None the less, the lack of party discipline means that there are opportunities for pressure groups and social movements at least to establish their issues as part of a political agenda, and hence it acts as an encouragement for movements such as the NCR to become active in electoral politics.

Additionally, as Vile (1976: 59) has noted, the increased centralization of American government—seen, for example, in the increased spending of the federal government or the centripetal tendencies of the judiciary—has not been accompanied by a corresponding strengthening of national party organizations.

Given the important part they play in the rise of the new Christian right, it is worth spending a little time on the American phenomenon of the 'political action committee' or PAC. The PAC is both evidence and cause of the lack of party cohesion in American politics. PACs came into their own as an unintended consequence of the 1974 Federal Election Campaign Act (Thompson 1984; *Congressional Quarterly* 1982). For many years, legislators had been concerned about the rising costs of elections and the possibilities of rich individuals or corporations buying influence with successful candidates by funding their elections. The scandal of Richard Nixon's campaign finance which came to light during the Watergate hearings gave a further impetus to reform. Although the law is complex, the main feature is tight restriction of the sums which individuals may give to candidates in any election. However, limits on the sums which a political committee may give to a candidate are either looser or non-existent. In 1982, an individual could only give $1,000 but a PAC could give $5,000. While individuals were restricted to total contributions to all candidates of $25,000, there are no controls on the totals which PACs may spend. Furthermore, there are no restrictions on what PACs may spend in tangentially helping a candidate—by attacking his opponent, for example—providing the PAC is independent of the candidate.

There are various types of PACs: corporate, labour, and trade, but the most dramatic change to follow the 1974 law has been the rise of the ideological PAC. In the 1980 elections, six of the nine wealthiest PACs were 'non-connected ideological' PACs. The Congressional Club, initially simply a support organization for North Carolina Senator Jesse Helms, a leading NCR supporter, raised and spent the largest sum of money: almost $8 million dollars. Second was another organization closely associated with the NCR: the National Conservative Political Action Committee (NCPAC, pronounced 'nickpack'). The three non-rightist PACs in the top nine were the National Association of Realtors, the United Auto Workers, and the American Medical Association, and their combined fund-raising was less than that of the Congressional Club.

The activities of various PACs will be discussed later. Here it is enough to note that the rise of the PAC has been accompanied by an apparent reduction in the importance of the political parties, and has thus given greater encouragement for social movement organizations to consider taking an active part in elections.

Further consequences of the above differences between the British and American environments will be mentioned at appropriate points. The general point to be made here is that there are significant differences between Britain and America, not only in the relative strength of conservative Protestantism, but also in aspects of the political, administrative, and communications structures, which encouraged American fundamentalists to see some prospects of success from direct involvement in the political process. Prospects and success are, of course, not the same thing. It may well be that supporters of conservative socio-moral positions in Britain are, in the long run, better served by an élitist and paternalistic centralized system. But it is the genesis of the new Christian right which is being explained and the above observations are offered as part of that explanation.

IDENTIFYING THE ENEMY: SECULAR HUMANISM

A crucial part in the mobilization of any group of people to campaign for or against something is the construction of an appropriate ideology. This does not have to be a completely integrated political or religious philosophy; it need only be a reasonably coherent story about why what is wrong is wrong and what can be done about it. Generally what is important about such stories is that they present an identifiable cause of the problems of which potential recruits to the movement are perhaps only fleetingly or ambiguously aware.

The ideological work of constructing a cause (in both senses of the word) involves two elements. The first is a simplification of the world and the second is a fixing of responsibility. In some contexts very little social construction is required. It is not difficult for whites in South Africa to see themselves as threatened by blacks. There are so many blacks, and whites have so much more economic and political power, that any major improvement in the socio-economic or political position of blacks is bound to mean losses to whites. If whites in South Africa feel anxious about their future they do not have to work hard to find a simple identifiable cause.

The problem for both potential supporters and promoters of the new Christian right was to construct a cause of the many things which concerned them about their environment. The solution was 'secular humanism', which:

> Denies the deity of God, the inspiration of the Bible and the divinity of Jesus Christ.
> Denies the existence of the soul, life after death, salvation and heaven, damnation and hell.
> Denies the Biblical account of Creation.
> Believes that there are no absolutes, no right, no wrong—that moral values are self-determined and situational. Do your own thing, 'as long as it does not harm anyone else'.
> Believes in the removal of distinctive roles of male and female.
> Believes in sexual freedom between consenting individuals, regardless of age, including premarital sex, homosexuality, lesbianism and incest.
> Believes in the right to abortion, euthanasia (mercy killing), and suicide.
> Believes in equal distribution of America's wealth to reduce poverty and bring about equality.
> Believes in control of the environment, control of energy and its limitation.
> Believes in removal of American patriotism and the free enterprise system, disarmament, and the creation of a one-world socialistic government. (Pro-Family Forum n.d.)

This list of beliefs was culled from two obscure documents—the first and second Humanist Manifestos of 1933 and 1973—which had an extremely limited circulation, and, as far as one can tell, almost no influence at all.[10] Anyone who held any of the above beliefs was labelled a 'secular humanist' and thus located as a member of the group that was actively undermining the world of the conservative Protestant.

It seems clear to an outsider that there is a considerable amount of simplification going on in this construction of secular humanism. While there is no problem recognizing that some people actively promote some of the beliefs listed by the Pro-Family Forum, it is difficult to see adherence to them as identifying a coherent body of people in the way that, for example, 'fundamentalism' identifies a particular subsection of conservative Protestantism. Many of those who want more equal distribution of wealth and the end of free enterprise are not environmentalists and care little for the agenda of the women's movement. Very many homosexuals are well-paid, high-status, fully supportive members of capitalist enterprises and would wish it no different. Many of those who oppose the conservative Protestant views listed in the first two items do so because they are adherents of competing supernaturalist belief-systems rather than because they are secularists, and so on.

An important role in the NCR's construction of secular humanisn is played by a footnote to the 1961 Supreme Court judgement in *Torcaso* v. *Watkins*, which said: 'Among religions in this country which do not teach what would generally be considered a belief in the existence of God are Buddhism, Taoism, Ethical Culture, Secular Humanism and others' (in Bollier 1982: 100). The appearance of 'Secular Humanism' in this list is thought to have been a reference to an actual church in California which called itself 'Secular Humanist'. Despite the poverty of this source, fundamentalists have used it to argue that the Supreme Court has declared secular humanism a religion. As will become clear in a later discussion of recent court cases, it is important to conservative Protestants that the views of the various groups which oppose them be describable as a 'religion'; such a rhetoric allows them to argue that a strict separation of church and state would outlaw the above

[10] If one totals the claimed memberships of avowedly humanist organizations, including Unitarian Universalists, one gets fewer than 200,000. The circulation of humanist pedriodicals is less than 22,000. These figures are given by Hunter (1986) arguing that secular humanism is a religion and they might be assumed to present the strongest case for the influence of humanism.

beliefs from public terrain as firmly as it outlaws conservative Protestantism. But the Court reference also serves the purpose of providing additional evidence for the claims that secular humanism exists and is actually promoted by an identifiable group.

In the final chapter I will return to the nature of social forces which threaten fundamentalism and argue that their amorphous nature explains why the new Christian right is unlikely to succeed in realizing its grand vision. Here I want to distinguish my observations about the usefulness of constructing a wide range of complex forces into an identifiable enemy from the Hofstadter and Lipset approach. They are undoubtedly right to see 'monism'— turning the complex into the simple and seeing unintended consequences as the result of a deliberate, preferably conspiratorial, campaign— as a common feature of extremist and populist movements. They are wrong both to exaggerate this feature and to regard it as a product of the social position of those who think this way. Uncertainty, complexity, and ambiguity may be features of correct thought, if by that one means the ratiocinations of 'thoughtful' people but they have never been features of ideologies which mobilize and motivate collective action, even when it is of the institutional (as in the case of stable political parties) as opposed to the uninstitutional sort.

The second caution which needs to be applied to any deployment of Hofstadter's 'paranoid style' notion is the appreciation of the way in which the motifs of styles of thought in a particular culture may predispose people to think in a monistic fashion. To a liberal Christian or an atheist, it may seem obvious that conspiracy thinking is so clearly defective as to need explaining, and that such an explanation must lie in the social strains endemic in the status position of those who espouse it. But consider for a moment the basic tenets of orthodox Christianity. There is a creator God who made the world and whose providence explains all the apparently inexplicable vicissitudes of life on this earth. There is also a devil. The world has order. There is an obvious continuity between the way that conservative Christians find order and cohesion in the complexities that surround them and the monistic style of thought which underlies the assumption that the world is the way it is because a conspiracy of secular humanists is deliberately perverting it. To summarize these observations, like Hofstadter and Lipset I want to indentify a discrepancy between the actual social forces which threaten the world of fundamentalists and their perceptions of that threat, but I do not want to conceptualize the thinking of those who believe they are threatened by a conspiracy of secular humanists as being radically different from styles of thought found in 'institutionalized' patterns of social action. Nor do I want to overlook a cultural explanation for them thinking like that. Where others see conspiracy thinking as a reasoning disorder brought on by status anxiety or structural strain, I see it as being at least partly a simple continuation of a style of thought which lies at the heart of most traditional religions.

However, the above points are secondary to the point on which I would agree entirely with Hofstadter, Lipset, and others; the mobilization of large numbers of people to collective action involves a simplification and dramatization of the causes of their discomfort. For the NCR, this was done by the discovery of the common feature of most of the things they did not like: the absence of consideration for, or reference to, traditional Christian beliefs and dogmas. This shared 'secularity' and 'humanism' was then constructed into the appearance of a coherent belief-system and imputed to those who promoted particular ideals or patterns of behaviour which the NCR rejected. Secular humanism provided an explanation for the many and varied changes which fundamentalists did not like and became the enemy against which they could fight.

CONCLUSION

To summarize these last two chapters, I have attempted to show: (a) why there should have been a population of conservative Protestants who had managed for a long time to preserve their culture and who had come to feel that the continued preservation of what they held dear required some active response to increased secularism, permissiveness, and government intervention; (b) how those concerns were stimulated and mobilized; (c) what features of American public life made direct political involvement appear to be possible and worth while; and (d) what was required by way of an explanatory and mobilizing ideology.

4

Chronicles: Nature and Actions

> *Moral Majority Inc. is made up of millions of Americans, including ministers, priests and rabbis, who are deeply concerned about the moral decline of our nation, and who are sick and tired of the way many amoral and secular humanists and other liberals are destroying the traditional family and moral values on which our nation was built.*
>
> Moral Majority brochure *c.* 1983

THE new Christian right is a general social movement rather than a specific social movement organization. Even organizational sub-units, such as Moral Majority Inc., are loose federations and alliances of groups of local activists. Hence it is not readily possible to list a series of aims and aspirations as *the* goals of the new Christian right. None the less, a cluster of desires can be described, although there is no suggestion that all who might sensibly be described as new Christian right supporters equally desire each of these things. Nor is it suggested that a clear line divides specifically NCR goals from the aims of other conservatives. New Christian rightists are against abortion and pro-traditional family, by which they mean that, while they would not deny basic civil rights to homosexuals (actually, some would; in a number of states buggery remains a criminal offence), they do not accept homosexuality as a legitimate alternative life-style. They want to maintain the traditional sexual division of labour and hence they oppose ERA (the constitutional amendment which would have guaranteed equal rights for women). They want to reverse the Supreme Court judgement which made it unconstitutional for public schools to sponsor organized prayer. They want changes in the tax laws to make it less expensive for parents to send their children to private Christian schools. They want school books to be vetted so that they do not promote 'secular humanism' and they want the Genesis story of the Old Testament to be taught as a plausible alternative to evolution as an account of the origins of human life. In addition, the NCR shares the secular conservative desire for heavy defence spending, an aggressive foreign policy, the unrestricted right to own guns, a curbing of trade unions, the dismantling of business legislation, a reduction in welfare spending, and an end to affirmative action programmes for ethnic minorities; in brief, either less government or a more conservative government.

THE NCR IN ACTION

Any discussion of the NCR has to bear in mind that there are really two distinct movements: federal and local. The point has already been made that the Moral Majority Inc., for example, brought together in one organization a number of groups which had already been engaged in socio-moral campaigns at city, county, and state level. If pressed too hard, the distinction is slightly misleading in that even the national campaigns, although orchestrated from the centre, are carried out by local groups. Later, in considering the overall impact of the NCR, attention will be paid to local issue campaigns but first those activities which are centrally led or which are common to most local groups will be discussed.

The Moral Majority and other NCR organizations work in much the same way as any other political campaigning movement. Although other things, such as lobbying legislators, are done, the main purpose is to get voters registered, educated, and to the polls. All other elements depend on being able to mobilize and direct a significant body of voters. Members of legislatures in democratic countries act in the way they do because they think there are votes in it. Unless the NCR can demonstrate that it can turn out votes, its agenda will be ignored.

NCR organizations know who makes up their constituency. Rather than trying to convert liberals to conservative positions, they sensitize conservative Christians to political issues and to their need to get involved. America is portrayed as a society which was great while it obeyed God's commands but which is being undermined by Godless 'secular humanists' who control Washington, the big corporations, and the media. America will become great again when the liberals and atheists are driven out of the Temple. This is the message which Falwell and other right-wing evangelists offer from their television pulpits and their direct-mailing computers. The NCR also seeks to recruit pastors to communicate this vision to their congregations. They are invited to rallies where the need for political involvement is explained, they are encouraged to preach on the need for activism from their pulpits, and they are taught how to take such practical action as organizing voter registration. This is vital. In Britain everyone is registered to vote. In America one has to go out of one's way to call at a certain office at certain times of the day. In the South the rules were designed to discourage blacks from voting but, until the late 1970s, fundamentalists had a similarly low rate of registration. Registering voters, preparing newsletters, influencing the media, contributing to candidates, and lobbying are among the skills which NCR organizations aim to teach through publications such as *The Christian's Political Action Manual* (Billings 1980).

Another part of the NCR campaign has been to 'target' liberal candidates in elections and make public their records on votes which can be construed as 'moral' or 'family' issues. Those legislators who have voted the wrong way on school prayer, funding for abortion clinics, the ERA, defence spending, the renegotiation of the Panama Canal treaty, and gun control are portrayed as being anti-family, anti-America, and anti-God. An example is the 1982 report card for Paul Sarbanes. At the bottom of the front page, under the Christian Voice title and logo is: 'When the righteous are in authority, the people rejoice: but when the wicked beareth rule, the people mourn. Proverbs. 29: 2.' On one inside page is the message:

> America was founded by men of faith on Biblical principles! Virtually all of our founding fathers recognized the crucial importance of religious morality as the foundation for our liberty and social well-being . . . Unfortunately, our Christian values and morality are today being drastically undermined . . . Not only is it our civic duty but, more importantly, our Biblical imperative to be God's standard bearer in the affairs of our nation. We cannot, in other words, rely on the unrighteous to safeguard morality in our government!

The facing page lists a series of issues on which Sarbanes voted. For each a brief title is followed by a short explanation and two columns, one which is blank, headed 'Your vote' and another which shows how Sarbanes voted. So the fourth item is 'For Parental Consent. Helms' amendment to require parental permission before providing sex education to children in schools. Yes = for parental consent'. Having read this, one is supposed to put 'yes' in the 'Your Vote' column and then notice that Sarbanes voted 'No'. After going through a similar process a number of times, the conservative reader should come to the conclusion stated at the bottom: 'Paul Sarbanes deserves our prayers but not our votes.' The back page explains the nature of the Christian Voice Moral Government Fund and invites people to write in for further information. This is standard; the pamphlets are widely distributed in every election and the write-in section allows Christian Voice to add to its mailing list of supporters.

Report cards were not an NCR invention. Liberal PACs had been issuing them for years. Since the Second World War, *New Republic* had issued 'liberalism' voting records (Turner 1970: 113). What caught public attention about the Christian Voice cards was that they evaluated legislators' views on many issues that had long been regarded as matters for private conscience and not for public debate. They also included voting on some issues that did not seem especially closely related to morality: the Panama Canal treaty, for example. Finally, the obvious tension between the 'morality' claimed for the ratings and the demeanour of some of those rated high (and low) created press interest. Most of the clergymen in Congress were given very low Christian Voice morality ratings. In contrast, one legislator whose voting had earned him full marks was to appear on FBI films of the 'Abscam' anti-corruption investigation, stuffing wads of bribe money down his shirt. Another 100 per center—Robert Bauman—was defeated in the 1980 elections after being arrested on charges of homosexual procurement. While it was quite reasonable of Gary Jarmin, the legislative director of Christian Voice, to argue that one could assess the morality of the consequences of people's voting independently of their personal morality, unsympathetic observers found the distinction difficult to accept.

In the earlier comparisons between the American and British political systems, the point was made that the NCR has been able to place its issues on the political agenda, even if it has yet to succeed in passing any major pieces of legislation. There is a rather subtle advantage for the NCR in the American legislator's ability to initiate bills, even when they are doomed to fail. Because they cannot blame their actions on a party whip, American legislators are often reluctant to permit roll call votes—those in which the names of those voting for and against are recorded—because their votes can be offered to the electorate for judgement. Senator Jesse Helms of North Carolina has made a speciality of constructing amendments on socio-moral issues and attaching them to entirely unrelated bills so that his opponents are forced to record a vote which can later be presented to their constituents as proof that they are 'anti-family', 'anti-life', and 'anti-America' (Furguson 1986: 121–31). In this way, the small amount of support needed to initiate an amendment and force a vote can lay the foundation for later negative campaigning.

NCR organizations have also sought to affect election outcomes through television and newspaper advertising, and, in so doing, have made their own contribution to the popularization of 'sleaze' advertising. Although American political advertising has often spent more time on attacking the opponent than promoting the candidate, some NCR campaigns, especially those which attacked liberals for being pro-abortion, achieved new depths of vituperation.

In addition to seeking to influence elections, NCR organizations have been involved in providing resources (such as temporary staff members, expert testimony, and published briefing materials) for a variety of other local campaigns to press school boards to change curriculum, to persuade city councils to close down pornographers, to mobilize support for state legislative equal time bills which would make the teaching of evolution conditional on equal time for the creation account of the origins of species, and so on.

THE SUPPORT-BASE FOR THE NCR

Initially Falwell denied that Moral Majority Inc. was a political organization. Early Moral Majority literature insisted that the organization 'was not a political party' and distinguished it from other NCR organizations such as Christian Voice by insisting it did not endorse political candidates. However, given that the Moral Majority did endorse orientations in such a way as to leave voters in no doubt as to which candidates to support, the distinction was largely a rhetorical nicety. The tensions which such a position involved eventually led, in 1986, to Moral Majority Inc. broadening its declared range of purposes and relaunching itself as the Liberty Federation. For all its ambiguity, the initial apolitical posture did reveal the problems the NCR had in legitimating its action in a society in which religion and politics are supposed to remain separate.

Falwell also denied that the Moral Majority was a religious organization. It is certainly the case that its aims, although generated by a particular religous culture, were promoted as if they could be isolated from the base. By 'religious', Falwell meant 'pertaining to salvational knowledge' and it was his great hope that the Moral Majority would not be a religious organization for, if it were, it would recruit only Baptist fundamentalists. For reasons which will be made clear, the NCR needed to transcend any particular denominational identity.

The *leadership* of the NCR was certainly successful at transcending confessional and denominational divisions. Weyrich is Jewish, and Viguerie and Phillips are conservative Catholics, as is Phyllis Schlafly of the Eagle Forum and STOP ERA. Falwell, Bob Billings, and Tim La Haye are fundamentalists. To date, however, the grass-roots support for NCR activity is still predominantly southern evangelical and fundamentalist, although specifically anti-abortion campaigns attract considerable Catholic support. Although surveys suggest that perhaps half the pastors of the Southern Baptist Convention congregations sympathize with the NCR (Guth 1984), the strongest support for the movement comes from those Baptists loosely grouped in a number of self-consciously fundamentalist associations. A survey of reactions to the Moral Majority in a northern city showed that more than a third of the 'ardent supporters' had been raised in the South and a similar proportion were some variety of Baptist (Buell 1983: 22). Although by 1986, the Moral Majority had organized chapters in all fifty states, the earliest and largest chapters were in the South or South West.

Although the leaders of NCR organizations emphasize their support from Jews and Catholics, their desire for a broad-based movement is not reflected in popular support. In a survey of Moral Majority support in the Dallas–Fort Worth area, only 9 per cent of Catholics were positive towards the movement. Of fundamentalist Protestants, 29 per cent were supporters. Even moderate Protestants were more sympathetic (14 per cent) than Catholics (Shupe and Stacey 1982: 41).

This narrow denominational base is not surprising. The 'holy roller' gospel shows were a key element in the mobilization of support. Moral Majority Inc. was launched with funds raised by an appeal mailed to all those viewers who had given funds to Falwell's *Old Time Gospel Hour* and, as the Liebman data referred to in the previous chapter demonstrate, many of the organization's cadres were recruited through Falwell's contacts in the Baptist Bible Fellowship and the Southwide Baptist Fellowship. In 1984, these two associations were joined by the World Baptist Fellowship and the General Association of Regular Baptist Churches in an Easter convention in Washington which pressed the government for more conservative moral legislation.

There are considerable obstacles to non-Protestant conservatives becoming involved. Considering value orientations first, Catholics share the NCR position on abortion, tax relief for Christian schools, opposition to 'gay rights', and ERA, but many have not forgotten that some NCR activists had, and perhaps still have, links with right-wing organizations such as the once anti-Catholic Ku-Klux-Klan, or that, at the end of the last century, fundamentalists were vocal opponents of separate Catholic schools.

Conservative Jews are in a similar position. Although the present NCR leadership has given considerable public support to the state of Israel, it is well known that such support does not result from any particular commitment to Zionism or from sympathy for Jews as such. Rather it comes from an eschatology—beliefs about the way in which the world will end—known as dispensationalism. Popularized by the over 18 million sales of Hal Lindsey's *Late Great Planet Earth*, this modern scenario for the end of the world believes that an Arab–African confederacy headed by Egypt will invade Israel. Russia and her allies will also attack Israel before turning on the Arab–African forces and destroying them. The Russians will kill millions of Jews before God kills five-sixths of the Russians. With the Russians out of the way, the final all-out war between Western civilization, led by the anti-Christ, and the vast hordes of the Orient led by the Red Chinese will begin. And thus the world ends. Before all this nastiness erupts, born again Christians will be taken up from the world in a 'rapture'. Their role of preaching the gospel will be taken over by the 144,000 Jews who will convert to evangelical Protestantism. At the climax of the final war, Christ will return to earth with the raptured saints, destroy all the bad guys, and the millennium—the one thousand years of righteousness—will begin. Thus although there is considerable

NCR support for Israel and the Jews, it stems from the part Jews will play in fulfilling scripture after they have given up being Jewish.

Furthermore, many American fundamentalists and right-wing political activists have a history of anti-Semitism, and Shupe and Stacey's Dallas–Fort Worth survey claims to find evidence that supporters of the Moral Majority are still significantly more likely to agree with anti-Semitic sentiments than are detractors (1982: 45).

The resonances of the politics of late nineteenth- and early twentieth-century conservative Protestantism are even more problematic for blacks. There are very large numbers of black Baptists and pentecostalists who are conservative in theology and on many socio-moral issues, and who should be natural recruits for the NCR, but who are conscious of the racist pasts of many of the people associated with NCR organizations. Senator Strom Thurmond, a grand old man of the NCR, first came to prominence in 1948 when, as the Dixiecrat presidential candidate, he represented white supremacist rejection of what now appear to have been rather feeble moves towards racial integration. Although Thurmond has since shifted his position considerably, Jesse Helms has not. While never as openly obstructive as Governor George Wallace of Alabama, Helms has always found good reason to oppose anything which might promote civil rights. His proposed solution to the school prayer controversy is to permit individual states to make their own decisions. At a press conference in 1983, he used the phrase 'states' rights': the old code for the southern resistance to civil rights legislation. Helms was the Senate leader of the opposition to the establishment of a national holiday to commemorate the birthday of Martin Luther King. After a very detailed and cautious examination of Helms's record, Furguson concludes:

> All his public life, he has done and said things offensive to blacks, and to anyone sensitive to racial nuance. He does them, says them, and eventually people complain. He either denies his action or its racial intent. But he never says he is sorry. Instead, he turns round and does it again, and again. Then he is called on it again, and the process is repeated, over and over. Its effect is to assure those whose base motive is race that while he has to be polite and deny those accusations in public, down deep he is one of them. It is one reason why Helms has won every election he has ever run. (1986: 231)

The American Independence party, to which Viguerie offered his services, nominated as its 1976 presidential candidate ex-Governor Lester 'Axehandle' Maddox, whose distinction, according to one of the leading intellectuals of the right, William F. Buckley, 'lies in his expressed preference for hitting a Negro over the head with an axe rather than serve him a plate of fried chicken' (in Crawford 1980: 237).

In addition, a number of NCR patrons and activists have been, or still are, associated with the anti-communist John Birch Society. Although the JBS has rarely been overtly racist, it denounced the civil rights campaign as a communist plot and has always contended that blacks suffer few disadvantages. Given this history of associations, it is not suprising that black Baptists, who would be happy to have public prayer in schools, are unwilling to work to achieve that goal in alliance with conservative whites they suspect of being at heart racist.[1]

Even when Catholics find themselves in agreement with NCR positions, as they do over abortion, there is a problem of organizational control. Catholic leaders, especially priests, are under the authority of the hierarchy, which has so far refused to endorse or give open support to the NCR. A similar point can be made about the Mormons. Senator Orrin Hatch of Utah has been active in promoting slightly weakened versions of NCR legislation but the Mormon leadership has not yet fully embraced the NCR, although the arrival in 1985 of Ezra Taft Benson as 13th President of the Church may herald greater official backing. Hatch himself is viewed as something of an opportunist by 'movement conservatives' and the NCR does not count on his support.

But even without these specific organizational or value-orientational obstacles, the NCR has limited appeal to anyone outside the WASP world because its arcadia—the one nation under God which was blessed—is white and Protestant. There are no parts for Catholics, Jews, or blacks in the sacred history of the new Christian right.

[1] Although some of the new right activists were previously involved in racist and anti-Semitic movements, the majority of those with previous rightist organizational involvement had worked with anti-communist groups such as Billy James Hargis's Christian Crusade and the John Birch Society, both of which are described in detail in Forster and Epstein (1964). Biographical details of conservative activists and details of the affiliations of new conservative organizations are regularly given in *Group Research Reports*.

THE LIMITS OF REGIONALISM

As the problems which fundamentalists have in broadening the support-base for their politics figure large in this account of the new Christian right, it is worth spelling out why the leaders of the NCR are prepared to work in alliance with various groups which they might normally be expected to oppose.

Fundamentalists need to recruit Catholics, Mormons, Jews, and secular conservatives because, without such support, they are powerless beyond the boundaries of their own regions. If, for the sake of simplicity, we take fundamentalists and evangelicals to be subgroups of conservative Protestants, some idea of their size can be gained from the following rough statistical description of American religious identification. In 1984, 57 per cent were Protestant, 28 per cent were Catholic, 2 per cent were Jewish, and 13 percent were 'other' or 'none' (Gallup 1985). The breakdown of denominational preferences among Protestants suggests that between 15 and 20 per cent belonged to a 'conservative' denomination in 1984. Thus even if all conservative Protestants shared the same political aspirations (which they do not) and acted as a block, they would still need the support of others outside that block to become a majority of anything.

Furthermore, they are concentrated in particular areas (Carroll, Johnson, and Marty 1979). Although the federal political system permits the persistence of regionalism, it mutes the influence of the strongest regions at federal level. This is especially true of the Senate in which each state has two seats irrespective of size. The small New England states have the same representation in the Senate as the large southern states in which conservative Protestants are more numerous.

Furthermore, the sorts of things which concern the NCR are issues so fundamental that they will normally have to be judged by the Supreme Court. Thus even when they command a majority in their own states and counties, conservative Protestants can be overruled. Although the Supreme Court is not oblivious to public opinion, it is only such major shifts as can be recognized in, for example, the passage of legislation in Congress, which would lead the Court to consider a significant change in its position on any of the issues which concern the NCR. Essentially, the fundamentalists can only effect the changes they desire if they are politically successful at federal level and that requires that they form alliances with other conservative groups.

5
Numbers: Impact and Influence

THE NCR was not slow to assert its own importance. In the 1980 elections, twenty-seven congressional liberals were 'targeted' and twenty-three of them were defeated, among them such notable figures as George McGovern, Frank Church, and Birch Bayh. NCR organizations claimed much of the credit for those results. They blamed the Republican party's failure to address their agenda in 1982 for the poor conservative showing in those mid-term elections. When Reagan won re-election with a 'landslide' in 1984, conservative Christians pointed out that he had gone out of his way to cultivate the fundamentalist vote by giving leading Baptist pastors—Jerry Falwell, W. A. Criswell, and James Robison—pride of place in the Dallas nominating convention, and by allowing a number of the NCR's legislative items into the party's platform.

But one has to be circumspect in assessing claims of impact. All the parties involved have interests and not all of them are obvious. For example, a number of conservatives who defeated targeted liberals in 1980 went out of their way to deny that the NCR had been an influence in their victories: a quite sensible reaction for politicians willing to accept any fruits of NCR activity without wishing to appear beholden. Some liberals wish to down play the importance of the NCR in order to prevent the movement acquiring further credibility. But some liberal organizations, People for the American Way for example, need to amplify the threat of the NCR in order to solicit funds to continue to campaign against it.

Before considering various indices of NCR importance, it is worth repeating the point that it is not easy or even sometimes possible to distinguish the NCR from the broader 'new right'. As a movement, the NCR consists of around twenty organizations with overlapping membership, leadership, and general support. On many issues, the main difference is in emphasis. The NCR shares the broader new right's desire for a more aggressive anti-communist foreign policy, more military spending, less central government interference, less welfare spending, and fewer restraints on free enterprise, but stresses social, moral, and religious issues such as 'the right to life', the reassertion of the male-dominated nuclear family, the reintroduction of public prayer in schools, the balancing of evolution by creation science in biology classes, and tax-breaks for independent Christian schools. It is difficult to know whether or not an organization such as the late Terry Dolan's NCPAC should be regarded as part of the NCR. Dolan himself is on record as saying that 'the government shouldn't be concerned with trying to wipe out sin' (Conway and Siegelman 1982: 98) and seems to have leant towards a libertarian indifference to legislation on socio-moral issues. When he died at the age of thirty-nine, he was widely reported to have had the AIDS virus (the very thing which Falwell described as 'God's judgement on homosexuals'). None the less, NCPAC has been quite happy to use the abortion issue in its negative campaigning against liberals. There are similar problems in placing the various organizations headed by Senator Jesse Helms of North Carolina. Their main interest (apart from advancing the career of Helms) is the promotion of the general conservative movement but as part and parcel of that goal, they have played an important role in the advancement of NCR candidates and issues.

Put briefly, the NCR overlaps with a wider conservative movement and many measures of influence and impact will relate to the wider movement. However, such confusion simply reflects the complexities of the motives and movements which constitute social action and, while we may wish it otherwise, the social scientist has to research and explain the actual world and not some simpler model of his or her own design.

There are a number of ways in which the NCR may have been politically influential. Before looking in detail at these, it might be noted that much of the data has the weakness that it was originally collected for some purpose (often as part of recurrent omnibus surveys) other than the evaluation of the impact of the NCR. Consideration of the influence of the NCR will begin a discussion of the progress of the NCR legislative agenda before going on to consider data on impact on federal elections. Finally I will consider state and other local campaigns.

THE LEGISLATIVE AGENDA

The clearest and most dramatic NCR success actually pre-dates the emergence of the new Christian right proper. Until a variety of conservative Christian organizations began to campaign against it, it seemed that the constitutional amendment to guarantee 'equal rights for women' when it was approved by Congress in March 1972 was well on its way to being ratified by the three-quarters of the state legislatures required. By the end of 1973, thirty of the thirty-eight states needed had accepted ERA. Then, under the leadership of conservative Catholic Phyllis Schlafly, sufficient opposition was raised through intensive letter-writing, petitioning, and phone-in campaigns to halt the bandwagon and then to persuade a number of states which had endorsed the amendment to reverse their decision. Only three more states joined in 1974, one in 1975, none in 1976, and one in 1977. While ERA was still one short of the three-quarters needed, Nebraska, Tennessee, Kentucky, Indiana, and South Dakota reversed their previous endorsements. Liberals pushed Congress to extend the time limit for ratification but this only gave the conservatives opportunity to persuade even more states to switch from support to opposition (Pines 1982: 169–71).

As it is an important point in assessing the impact of the NCR, it is worth noting that the anti-ERA campaign, although it was directed at a national level, and concerned what was initially and ultimately a federal issue, was in operation a coalition of local efforts.

In terms of federal legislation, the NCR has made almost no progress at all on its agenda. Most bills promoted by NCR supporters have died in committee. Even on relatively simple matters such as school prayer there has not been enough momentum to force the authors of different bills to combine them into a single measure which might have won some support. The most popular item on the agenda—the outlawing of abortion—has made no progress. Hatch's 1983 constitutional amendment failed by eighteen votes to reach the two-thirds majority and failed by two votes to muster a simple majority. Helms's bill to make abortion illegal, introduced in the same session, was successfully filibustered. The very few policy decisions which the NCR could claim as supportive of their positions (tightening the availibility of federal funds for abortion referral clinics, for example) were taken by relatively low level functionaries within the Reagan Administration.[1] The NCR simply has not had enough support in the Senate (and far less in the House of Representatives which, even in the most conservative years, from 1980 to 1986, had a large Democratic majority) to pass its bills.

However, the NCR can claim some very minor strategic victories in the already mentioned work of Jesse Helms in forcing roll call votes on socio-moral issues. With no hope whatsoever of having any of his measures passed, Helms has been able to use his excellent parliamentary skills to embarrass liberal senators by forcing votes on measures and amendments which were carefully worded so that they sounded like the sort of thing any decent person should vote for, but which moderate liberals felt obliged to oppose. Votes against the amendments were later used against liberal incumbents when they stood for re-election.

The legislative failure in the federal Congress has largely been mirrored by a similar failure in state legislatures where countless NCR-sponsored measures have been allowed to expire quietly at the committee stage. However, in Arkansas and Louisiana bills were passed which obliged biology teachers to give equal time to 'creation science' as a valid alternative to evolution as an explanation of the origin of species. More will be said on these measures later.

The general legislative failure does not surprise nor disappoint the professional activists in the NCR. When I first interviewed a number in 1983 they saw the bills they then sponsored as markers for the future and quite sensibly pointed out that real progress required the building of conservative majorities in both chambers of Congress, something which would depend on enduring and increasing electoral impact.

ELECTORAL IMPACT: VOTER REGISTRATION

If the NCR was to have any effect, its first task was to overcome the traditional pietism of southern fundamentalists who had long held

[1] In assessing the impact of the NCR, it is worth noting that Jo Ann Gasper, the official responsible for one of the few Administration decisions which the NCR could claim as a success— restricting state funding to some pro-abortion groups— was subsequently forced out of the Department of Health and Human Services, apparently for exceeding her remit (*Group Research Report* June 1987: 22).

the 'world' in disdain, as something to be avoided while one concentrated on the primary task of saving souls. Falwell himself had held and often expressed such a position in the late 1950s and 1960s. Having concluded that conservative Christians should become active, he began to advocate participation in the political process, and other NCR organizations such as Christian Voice and Religious Roundtable devoted considerable energy to persuading pastors of the need for involvement and in teaching them how to politicize their flocks. The main priority of the NCR was never to convert social and moral liberals into social and moral conservatives but to persuade those who already held the appropriate values to become politically involved. NCR organizations report considerable success in persuading people who had not previously registered to do so (and some of the data presented below on actual voting would seem to support such a claim). According to the *American Political Report* (30 November 1984), the American Coalition for Traditional Values (ACTV), an umbrella NCR organization which included the Moral Majority, registered 150,000 new voters in North Carolina for the 1984 election in which Helms fought off a strong challenge from a popular Governor. In the words of Lamarr Mooneyham, an independent fundamentalist pastor and head of the Moral Majority in North Carolina:

> We put one registrar in each church. We'd go anywhere anybody was interested. It was mostly just conservative churches, evangelical, fundamental, not necessarily Baptist. The common denominator was they were conservative in theology . . . We had a table there every Sunday, out in the hallways where the people would come out after services, or sometimes outdoors if it was nice. That's allowed. Under North Carolina law, a church is like any other place. (Furguson 1986: 169–70)

ELECTORAL IMPACT: VOTING

Registering people to vote is of little value unless they actually do so, and vote for candidates most sympathetic to NCR goals. It has to be admitted that evidence for the impact of NCR activity on voting behaviour is extremely difficult to assess. The initial claims for NCR importance, made when four liberal senators who had been targeted by new right groups were defeated in 1980, were quickly rejected by analysts such as Lipset and Raab (1981) who pointed out that the swing against the four liberal Democratic senators was much the same as the swing against Democrats in contests where the NCR had not been active.

Such scepticism is shared by Johnson and Tamney, who argue, on the basis of their recurrent survey of residents of Muncie, Indiana (the Lynds' 'Middletown'), that 'as for the Christian right . . . in the 1980 presidential election, it proved to have no significant impact at all' (1982: 130).

AN ASIDE ON METHODOLOGY

Before proceeding further, some cautionary notes should be sounded. What one finds in survey research depends on what one measures. I do not want to suggest that Johnson and Tamney have been lax in their research but I would like to point to one or two problems which affect many of the surveys reported in this discussion.

Firstly, research which is designed to discover information about people's attitudes, beliefs, and intentions is only useful if there is an intuitively plausible connection between the information being sought and the questions asked. No amount of sophisticated statistical analysis improves data gathered with bad questions. Furthermore, there are so many subtleties which can be represented by such analysis that demonstrating that a variety of statistical measures *cohere* (that is, that something is being represented) cannot, of itself, be grounds for assuming that such devices represent what they claim to represent if there are legitimate doubts about the initial questions. Statistical techniques cannot improve data.

As an example of my concerns about the relationship between questions and answers, and the 'data' which are required for the argument, I would mention the Johnson and Tamney 'political activism' variable. Four dimensions of support for the Christian right were used to measure political activism: political involvement, belief in 'civil religion', religious fundamentalism, and a desire for voluntary public prayer in schools. Political involvement was measured as the extent of agreement with the following statements: 'The country would be better off if there were a religious political party' and 'One's minister (priest, or rabbi) should come out in support of a presidential candidate.' To someone who was not well steeped in Protestant politics, such questions might seem sensible guides to religious political involvement. But those with any close acquaintance with the NCR know that most new Christian right activists would, at least in public, disagree with both statements. Conservative Protestants do not want a religious party: they want both secular parties to adopt socio-moral positions which are acceptable to conservative Protestants.[2] Most NCR activists accept the separation of church and state. It may seem to outsiders that the logical extension of NCR positions is a form of Christian politics which could or should be pursued most effectively by the formation of a religious party, but most people involved in the NCR wish to maintain an at least rhetorical commitment to the conventional democratic structures of a pluralist society. For similar and related reasons, few NCR activists I have interviewed would be happy about a pastor publicly supporting a candidate. Although Falwell now gives such endorsements, he did not do so in the early days of the Moral Majority. Many pastors operate a degree of compartmentalization. Preaching on the merits of Christian voting is laudable. Encouraging people to register to vote is a good thing. Expounding biblical positions on abortion or foreign policy is a pastor's duty. Giving extremely heavy-handed hints as to which candidate comes closest to the correct Christian position is also acceptable. But, and it is a big but, there is often reticence about the final step of actually endorsing a candidate.

The reticence is sometimes guided by the rather mean motive that endorsement may threaten the tax-exempt status of the Church. More often it results from a genuine concern for the possibly divisive effects of too blatant partisanship. Pastors who are generally supportive of the NCR, and who spend their leisure time working for NCR organizations and campaigns, none the less operate a division of labour. When they are being pastors, their job is to preach the word, not to tell people who to vote for in elections. One might suspect a certain disingenuity in such a position but it is the position of many NCR-sympathizing pastors.

More generally, it should be noted that many NCR activists follow the lead of Falwell and other leaders in attempting to frame their religiously produced concerns in a non-religious rhetoric. Thus creationism becomes creation science. Conway and Siegelman (1982: 110) note that anti-abortion campaigners 'dressed their

[2] Even the third-party notion in the form presently being canvassed by Jarmin and others, discussed in ch. 7, is explicitly not a Christian party.

cause in grandiloquent medical and legal terms, only admitting among each other that their cause remained a religious one'. Whether this represents a genuine attempt to transcend religious particularism or is simply a rhetoric adopted to increase the chances of legislative success is not important at the moment. What is important is that the two statements used by Tamney and Johnson as indicators of political involvement are too blunt; perhaps only fractionally too blunt, but, nevertheless, sufficiently insensitive to the nuances of the ways in which people perceive the links between religion and politics to miss the mark. To be fair to Tamney and Johnson, I do not think any other pair of short statements could have been better constructed. What we have here is simply the problem of finding short questions to which people can give answers which genuinely reflect their feelings, beliefs, and attitudes.

The objections to certain sorts of survey research can be taken to a higher level of abstraction and presented as a general critique of positivism. In a footnote to one of their papers on NCR support in Muncie, Indiana, Tamney and Johnson (1983: 151–2) point out that they have very few members of fundamentalist churches in their sample. The implications of what for Tamney and Johnson is merely a footnote point are actually so far reaching that they need to be spelled out clearly. We know that some people support the NCR. Some of them claim that their support of the NCR has an effect on their political decisions. To test the extent of NCR impact, Tamney and Johnson create an abstract cluster of 'variables'—NCR support—and search for traces of this among a representative sample of people. They announce their result—that NCR support, as measured by their variables, does not affect voting behaviour. But why should it affect the voting behaviour of non-NCR supporters? That some people are motivated to act in way *X* by a complex of beliefs *Y* does not mean that smaller traces of *Y* should cause a commensurately weaker tendency to act in way *X*. Hence the discovery that people with a weak allegiance to beliefs *Y* do not act in way *X* tells us little or nothing about the relationship between beliefs *Y* and actions *X* in the case of those people who are fully committed to beliefs *Y*. Tamney and Johnson are telling us stories about the influence of religious fundamentalism and three other NCR variables on the behaviour of people who are unlikely to be strongly committed to religious fundamentalism. This seems to be a good example of misplaced reification.

Fundamentalism is not a single property which is distributed universally in varying quantities so that one can describe liberals as 'people with zero fundamentalism'. It does not follow from the proposition that fundamentalism (*F*) causes action *Y* that $\frac{1}{2}F$ should cause $\frac{1}{2}Y$. Hence the demonstration that $\frac{1}{2}F$ does not cause $\frac{1}{2}Y$ leaves us no better informed about the truth value of the claim that *F* causes *Y*.

It is likely that the consequences of particular beliefs or attitudes vary in a non-linear fashion depending on the ideological or attitudinal packages in which they are held. It is also likely that certain types of belief influence motives in a manner similar to electronic 'gates', so that rather than having a regularly increasing impact, they have no impact at all until a point at which they 'trip' a response. If either or both of these are likely, then one has to suspect that investigations of the impact on voting of sympathy towards NCR positions, which search for a relationship amongst people whom we can anticipate will be unsympathetic to the NCR, are hardly to the point.

To summarize this objection in a single proposition, I have considerable reservations about research which attributes causal efficacy to characteristics which have been abstracted from the people who exemplify them. The borrowing from natural science of the idea of a 'variable' seems itself misleading for the explanation of social action. It is easy and convenient to shift from talking about why fundamentalists act in a certain way to talking about the causal efficacy of 'fundamentalism', but it may often be misleading, especially when one searches for the effects of the property 'fundamentalism' on the actions of people who are noticeably short of it. As Blumer (1967) ably argued in his criticisms of positivism, most varieties of social action involve complex processes of interpretation. One simply cannot sensibly treat people as if they were black boxes with measurable inputs (independent variables) and outputs (dependent variables). Inside the box is interpretation. Identifying 'independent variables', let alone using them to explain anything, requires a detailed knowledge and understanding of the interpretative processes employed by the actors. It might be argued that, even if we could know every relevant belief that is inside the head of an actor, we could not predict with certainty his or her actions or attitudes to any particular phenomenon. But even if one does not go that far, it should be obvious that the knowledge gained in many surveys falls very far short of fully describing any actor's stock of relevant knowledge.

There are also good reasons to be suspicious of research which attempts to treat attitudes as if they were, in varying degrees, independent of the particular occasions, plans, and schemes in which they might be expressed and discovered. An example can be given from research on attitudes towards ecumenism. Black (1986) reports a number of studies of members of Protestant denominations which found little correlation between theology and attitudes towards church union. As he sensibly points out at the end of his review, this is not surprising given that the studies treated 'attitudes towards church union' as if they were independent of the identity of the church with which the respondents were being invited to consider union! What is obvious from a moment's reflection is that whether church union is a good or bad thing depends on the similarity between the respondent's own religion and that of the other body in any proposed merger. For example, in the Church of England, low church evangelicals are more positively ecumenical towards the Methodists than they are towards Rome. High church Anglicans are happy to consider union with Rome and the Orthodox Churches but are opposed to union with Protestant dissenters. This is the second danger of insensitive attitude research; that in treating attitudes as 'variables' it may either completely divorce attitudes from their objects or attach them to some reified abstraction ('church union', for example) which does not actually play a significant part in people's reasoning.

The above comments represent only a small part of the reservations that a sociologist of a symbolic interactionist and Weberian bent has about the positivist assumptions on which most quantitative research rests. There are good theoretical and methodological grounds for entirely dismissing survey data of the sort employed by Johnson and Tamney. However, given that some of the problems are variable, each case will be assessed on its own merits. The nature of the issues being examined, the questions used to derive information, and the statistical procedures used to analyse the data all have implications for our assessment of the value of the final results.

To return to the substantive conclusions of the Johnson and Tamney research, further doubts about its merits are provoked by the ability of another analyst to use the same data to come to a quite different conclusion. Simpson argues that:

> The findings . . . clearly indicate that voters' orientations to socio-moral issues did have an effect that favored Reagan in both 1980 and 1984. . . . the NCR has had an impact on recent presidential elections but it was not the direct effect achieved through the delivery of the born-again Christian vote. Rather by successfully politicizing socio-moral issues, organizations such as the Moral Majority gave Reagan an opportunity to identify with the closely-held personal moral values of the majority of Americans and to engage in a politics of morality which favored him at the polls. (Simpson 1985: 121)

Although one may still be sceptical about the degree to which the survey data actually support that conclusion better than they support the Johnson and Tamney view, Simpson's discussion fits better with what one concludes from detailed interviews with conservative Protestants, and appears more likely to be compatible with the sort of understanding of the NCR which develops from close acquaintance with NCR supporters. The extent of the impact of the NCR is not reducible to the size of a solid constituency of conservative Protestants who 'join' the NCR and then, acting in accordance with its directives, vote for Reagan when they would not otherwise have done so. The world is not that simple.

Other studies support NCR claims that there has been a recent politicization of conservative Protestants. Smidt (1983) compares voter turn-out of southern and non-southern evangelicals and non-evangelicals. Using University of Michigan Centre for Political Studies data, she reports the following degree of politicization of the various groups shown in Table 2. The implications are clear.

TABLE 2 *Politicization of White Non-evangelicals (NE) and Evangelicals (E) by Region as a percentage*

	South		Non-South	
	NE	E	NE	E
Voting turn-out in prior pres. elections				
All/most	70.5	61.1	73.2	60.8
Some/none	29.5	38.9	26.8	39.2
Voted in 1980	65.9	77.0	73.3	74.6

Prior to the 1980 election, evangelicals were considerably less likely than non-evangelicals to vote. In the South the participation rate of non-evangelicals actually declined while that of evangelicals rose by 16 per cent. The participation rate for non-evangelicals outside the South remained the same while that of evangelicals rose by 14 per cent. If one were to take only these data, it would be possible to suppose that the increased evangelical participation brought benefits for both Carter and Reagan but it is well established from a number of surveys that, although Carter benefited from southern conservative Protestant votes in 1976, in 1980 they went wholesale to Reagan (Phillips 1982: 189–92).

Another report on the Michigan data (Miller and Wattenberg 1984: 314) also shows a link between fundamentalism and voter turn-out. Eighty-two per cent of the 'most fundamentalist' respondents reported voting in the 1980 presidential election while only 71 per cent of the 'least fundamentalist' voted.

It is clear that groups such as the Moral Majority are not popular. In a survey of attitudes towards twenty-two categories of person, the Moral Majority was rated at only forty-five degrees on a 'feelings towards' thermometer, while the average for the twenty-two groups was sixty-seven. Even 'people on welfare' rated higher, and only 'radical students' and 'black militants' were less popular (Miller and Wattenberg 1984: 304). However, in determining election outcomes small numbers can be significant. After all, the difference between Reagan's result in 1980 and his 'landslide' victory in 1984 was the difference between 52 and 59 per cent of votes cast. Secondly, it might be that there is a constituency of people beyond the small group who actively support the Moral Majority whose voting behaviour is still affected by their conservative religious beliefs. Buell and Sigelman's (1985) analysis of the Michigan data suggests that many even of those people who agree with NCR positions are none the less ambivalent towards the Moral Majority.

Although they do not pursue the issue, this may well reflect the presence of Catholics and, another group not so far mentioned, 'Bob Jones University' fundamentalists. A substantial body of fundamentalist opinion is entirely at one with Falwell in his condemnation of the moral state of America but remains loath to abandon its position of pietistic disengagement. Another possibility has already been mentioned. Many conservative Protestants who are not ideologically committed to the Bob Jones University position continue to be ambivalent about voicing, even to themselves, support for a degree of religious influence on political action which might be seen as contradicting the widely accepted idea of a separation of church and state. Hence many of the conservative Protestants I interviewed preferred to see their socio-moral positions as having some generalized source, such as 'Americanism', other than, or larger than, their religious commitments.

Using a rather loose notion of fundamentalism, Miller and Wattenberg find a direct correlation between fundamentalism and voting in the 1980 presidential election. When voters are arranged on a scale of religious orthodoxy with six positions, while only 47 per cent of the 'least fundamentalist' voted for Reagan, 85 per cent of the 'most fundamentalist' did so. As one might expect, religiosity does not seem to have contributed to the voting decision, independent of its overlap with party identification and liberal or conservative political ideology. That is, although religiosity was implicated, in that there is a strong connection between fundamentalism and conservative political ideology, there were few cases of people who were liberal on socio-economic issues voting for Reagan because they were conservative in their religion and approved of his socio-moral positions.

However, Miller and Wattenberg do claim to find evidence that religiosity *did* have a significant and independent effect on voting for Senate and House candidates. Indeed, religiosity had a greater effect on these decisions than did liberal–conservative orientation. To put it in human terms, some identifiable people who held liberal views on socio-economic issues, but who were conservative in their religiosity, voted for the more conservative candidate.

ELECTORAL IMPACT: CAMPAIGN FINANCING

A very obvious NCR effect which does not appear in surveys which seek correlations between support for NCR values and voting for Reagan is the raising and distributing of campaign funds. It is possible that the NCR has helped conservative election candidates by channelling funds to finance their campaigns, and that such increased expenditure has allowed a conservative candidate to do better than he or she would otherwise have done. For example, in funding his 1984 re-election campaign Jesse Helms received considerable public support from the Moral Majority and the American Coalition for Traditional Values (ACTV). He also received money from these and other NCR organizations. In fact, one could almost describe Helms's Congressional Club political action committee as itself being an NCR organization. With regard to Falwell's endorsement and NCR money (a) both may have had considerable impact, (b) both may have had no impact, (c) only one of the two might have had any significant consequences, or (d) (and this is the complexity which defeats most secondary analyses of survey data) the consequences of funding and endorsement may have run in different directions. It is quite possible that Falwell *lost* Helms votes by his endorsement (if more people voted against Helms because they disliked Falwell than voted for him because they liked what the Moral Majority stood for) while at the same time *gaining* him votes by helping him to raise money from out-of-state conservatives or by helping him to extract more money from his safe supporters than he would otherwise have been able to do.

The point of this tortuous reasoning is to demonstrate that the NCR's ability to raise money and deploy it in election campaigns is a factor which has to be considered independently of the obvious effects of NCR propaganda on the voting intentions of individuals.

What is known is that NCR (and secular new right) organizations have been able to raise very large sums of money in small amounts from large numbers of people, largely through the use of direct mail. However, not much of this has gone directly to conservative candidates in elections. Or, to be more painfully precise, it *appears* that not much of this has gone to conservative candidates. In fact, even the most detailed research has found it difficult to discover funding details. Latus (1984: 251–3), for example, reports that the Christian Voice Moral Government Fund spent $425,565 for the 1982 elections and that only $12,500 was spent on behalf of candidates. According to Gary Jarmin, the administrator of the fund, a much larger proportion was spent to help candidates but was listed as 'administrative expenses' in the report to the Federal Election Commission because such a description saved Christian Voice a lot of office work.

If we accept that far less was spent in supporting conservative election candidates than was raised by NCR organizations, there may be a number of reasons for this. The fund-raising organizations are themselves expensive to run (or greedy, depending on one's view). The Viguerie Organization, for example, kept almost 50 per cent of the funds it raised in one campaign. In raising $802,028 for Bibles in Asia, the Viguerie company accrued expenses of $889,255, thus costing Bibles in Asia more than it raised (Crawford 1980: 63). Secondly, as Koenig and Boyce (1985) have shown (although they do not seem to appreciate the point of their own data), most conservative congressional candidates do not *need* NCR money because they are well funded by PACs representing more traditional conservative groups, such as business interests.

In order to evaluate the impact of NCR fund-raising, we need also consider whether NCR groups were *better* at fund-raising than their liberal counterparts. The University of Michigan data show no strong correlation between fundamentalism and making political donations but it has the weakness that it only compares the number of donations and not the totals involved (Miller and Wattenberg 1984: 314). If the difference between the amount given by fundamentalists and liberals to their churches is mirrored in their political donations, then NCR organizations would have acquired more than liberal PACs. On the other hand, it may be that most NCR supporters already give so much of their free income to church causes that they have little left for big donations to political causes. All that can be said is that the 'non-connected ideological' PACs which raised most money for recent elections have been of the right rather than the left. However, any clear conclusion is impossible, given that NCR supporters may have been mobilized to channel funds through a series of local organizations.

ELECTORAL IMPACT: NEGATIVE CAMPAIGNING

Koenig and Boyce (1985) take the fact that those congressional candidates who were strong supporters of the NCR agenda did not receive the bulk of their funding from NCR organizations as evidence of the ineffectiveness of the NCR. There are a number of reasons for being suspicious of this conclusion. The first is that most NCR organizations took the deliberate decision to let incumbents look after themselves. Incumbents have a high chance of being re-elected because they have access to a considerable range of resources which are denied to the challenger. They have almost endless opportunity for good media coverage; they can promote bills to benefit their constituents; and they have a large paid staff, the privilege of a number of free mailings, and access to a considerable 'pork barrel'. As one experienced lobbyist put the case to me: 'Any Senator or Representative who can't use his position to earn enough IOUs while he's in office is the sort of bozzo we can do without.' Thus, in so far as the NCR did channel funds into election campaigns, their effort would have been directed to a small number of contests, usually ones where a conservative was challenging a liberal incumbent.

In addition, the NCR spent most of its money on general sensitizing campaigns, voter registration drives, and other such activities which *prepare the ground* for the conservative candidate. This sort of work, while it may be a bonus for the conservative, would not show up as a direct asset.

Where the NCR did spend its money was in negative independent expenditure. 'Independent' means spent without the direct involvement of the beneficiary and independent expenditure acquired a new importance after the passage of the 1974 campaign law which attempted to control the amount of money which could be given to any particular candidate. The loophole is that there are no limits on spending which 'is done without consultation with the candidate it is designed to help' (*Congressional Quarterly* 1982: 80). The nature of independent expenditure can readily be discerned from the direction and size of the sums deployed in 1980. Less than half a million dollars was spent 'for' Democrats while 13 million was spent for Republicans. Just over $100,000 was spent 'against' Republicans but over 2 million dollars was spent opposing Democrats. The majority of that money was deployed against the six liberal senators targeted by the NCR. Four of the six were defeated but, as has been pointed out, we have no sure way of knowing if those defeats were a result of the independent expenditure campaigns and the negative advertising they funded.

Some observers claim to have seen a backfire effect in the NCR's efforts. This could have taken at least two forms. Waverers may have reacted to what they saw as unfair criticism of liberals by either shifting to those candidates or, if they were already 'weak' liberal supporters, by making a point of going out and voting. This seems to have happened in the 1986 mid-term Senate election in Maryland, where Linda Chavez, a strong supporter of new Christian right positions, lost a significant lead (and what had been a Republican seat) to Barbara Mikulski, a liberal trade-union activist. Commentators believe that there was a significant shift to Mikulski as a reaction to Chavez's barely disguised claims that Mikulski was a communist, child-murdering lesbian.

Other explanations of this result are possible. Every NCR loss can be, and has been, explained by local factors. However, received wisdom has an amplifying effect. In the early days of the NCR its 'successes' were attended to and it hardly seems coincidental that many liberal candidates became increasingly defensive about their records on social and moral issues in an attempt to protect themselves from the attacks of NCR negative campaigning. Latus suggests such a possibility in her account of usefulness of the National Christian Action Coalition's 'family issues voting index'. Like the Christian Voice score card, this shows the way in which incumbents voted on a variety of pieces of legislation on which the NCR has a position. The NCAC index was widely used in a number of campaigns in 1980.

> The index was not particularly helpful in 1982, however. First, there were few votes on the sort of family issues that concern Billings's group. Second, in the few races where candidates were rated, legislators seemed to have repented of their antifamily sins. Because they 'voted right' on these issues there were fewer low scorers on the index, reducing its political usefulness. (Latus 1984: 254–5)

Thus even if the campaigns did not work in any obvious objective sense, they may still have had the consequence of undermining the confidence of liberals and creating a more conservative climate.

If it is the case that public, and especially press, reaction can exaggerate small shifts in voting behaviour—the case with the early years of the NCR—then the process should work in reverse at the first signs of NCR reverses. The failure of the NCR to hold some seats and to win others in 1986 opened the way for the construction of a 'death of the NCR' story. This point will be taken up again later in a general concluding assessment of the NCR.

NON-ELECTORAL IMPACT

Although the NCR's involvements in federal elections and national campaigns are the most visible fruits of the movement, there is a good case to be made for arguing that it is in local small-scale struggles that NCR organizations and their supporters have had their greatest impact. Many good examples can be found in conflicts over schooling.

The general structural condition which makes such campaigns possible is the presence of an elected element in the administration of public schools. Professional educators in America (and their associations, such as the National Education Association) tend to be liberal and cosmopolitan, as they are in other advanced industrial societies. In the past, elected school boards have often endorsed the actions of the professionals, largely because when education is not an issue there is little interest in the elections and places are filled by people with values and interests similar to those of the professional educators. Additionally, there is always the imbalance of authority, with elected representatives tending to defer to experts. However, the strong rhetorical stress on local democracy in America always makes it possible for parents and other concerned members of the community to intervene in schooling policy.

The most spectacular recent intervention is worth examining in detail because it illustrates both the ideological content of recent disputes and the structural conditions which allow disputes to develop. Kanawha County, West Virginia, is a part of Appalachia in which the population is divided between the coal miners of the mountain region and the educated white-collar workers of the town of Charleston. In 1970, Alice Moore, the wife of a fundamentalist Church of Christ minister, was elected to the local school board. In 1973 she organized a successful drive to ban sex education. By 1974 she had become concerned about the 'secular humanist slant' of textbooks being used in local schools. As she put it:

> They used mythology throughout and compared various myths to Bible stories, implying that the Bible was a myth. They were very critical of parents, very critical of authority in general. Some of it was very good literature, but the editors had selected the stories in such a way that you never saw a Christian who wasn't a hypocrite. Atheism was presented always intelligently, always attractively. (in Dewar 1983)

Moore decided to campaign against the textbook selection and toured the area, speaking to local church groups. At one meeting:

> About 60 people showed up . . . and after they'd seen the books, they wanted to do something. They started circulating petitions. Various ministers started calling me up over the summer and asking me to come

speak at their churches. I just told them what was in the books. They took it from there. By the time school opened there was a lot of tension, a lot of anger. People had no outlet. They were being told that their children were going to use these books and there was nothing they could do about it, so they did the only thing they could do. (in Dewar 1983)

They withdrew their children, picketed school board meetings, and blockaded schools. As feelings intensified, the protests spread from parents to the mining community at large. The coal miners came out on strike to support the parents.

Moore put the issues clearly:

What we are fighting for is simply who is going to have control over the schools, the parents and the taxpayers and the people who live here or the educational specialists, the administrators, the people from other places who have been trying to tell us what is best for our children. We think we are competent to make those decisions for ourselves. (in Crawford 1980: 156–7)

In his discussions of nativist movements, Lipset makes an interesting case for the operation of a parallel to Gresham's law of good and bad money. In protest movements, the initial quite moderate, more 'middle-class' support is displaced by a rowdier, more aggressive element as the temperature of the dispute rises. This certainly was the case in Kanawha County. The initiative began to move away from the ministerial leadership. The Ku-Klux-Klan joined in demonstrations. Elmer Fike was a local business man who orchestrated the early protests and later became the head of the Moral Majority in West Virginia (incidentally, another example of the way in which the Moral Majority Inc. mobilized on the foundations of earlier local campaigns). Looking back on what he now sees as a 'noble failure', he said:

The thing that broke the back of the texbook protest was the fact that we had several strong-willed people who didn't want to submit to any kind of discipline at all. A lot of these people had never been in the public eye, and they were basking in the glory and the national publicity and they just wanted it to go on forever. The crazies were out . . . and some of them were really wild. (in Dewar 1983)

They were wild enough to firebomb three schools and to shoot at teachers and officials who had been targeted by the protesters.

Interestingly, given the general liberal impression that such parents are dangerous reactionaries motivated by irrational and anxious responses to any sort of change, the liberal *Nation* magazine accepted that:

> white working-class parents from coal fields and Southern mountains do have legitimate grounds for complaint. The editing reflects a value system that runs counter to most of what they cherish. The supplementary books, in particular, play out the alienation felt by urban intellectuals and university militants of the 1960s, who seem to the protestors to have taken over the publishing houses in New York. Some of the selections were unpatriotic, sacrilegious and pro-minorities, and they would, as the parents predicted, legitimize different values and raise heretofore taboo questions. Equally important was the almost total exclusion of people like themselves from the 'multi-cultural' texts. The editors had not thought that coal miners, or country folk, or Appalachians or working people were either distinguishable groups or important enough to include. (quoted in Crawford 1980: 157)

Why Elmer Fike now sees the protest as a failure is not clear. Although the conservatives were embarrassed by the imprisonment of a fundamentalist pastor for firebombing (a charge he always denied), the most controversial books were withdrawn and in subsequent elections Moore's group took almost all the seats on the school board. Although they did not abandon control to the parents, the professional educators became considerably more cautious and sensitive to local values.

Although Moore found support for her campaign in the publications of professional fundamentalist educational organizations, the Kanawha County controversy was initiated by local people who had themselves come to the conclusion that the sorts of values being transmitted to their children were antithetical to their own. Since then professional agitators have become increasingly important in stimulating local disputes.

As an example of conflict stimulated by outsiders, there is the case of the small town of St David's, Arizona. In January 1982 an organizer for Schlafly's Eagle Forum arrived and announced a public meeting on the evil effects of secular humanism. The meeting was shown a film of scenes from New York's sleazy 42nd Street and convinced that such degradation would be the result of the sort of education their children were receiving. Under pressure from concerned parents, the local school board set up a committee to review its teachers and the books in the school library. The committee, briefed by the Eagle Forum, ended up focusing its discussions on one teacher who was denounced for disseminating secular humanism. William Golding's *Lord of the Flies* and books by Steinbeck, Conrad, and Twain were banned. At the end of the school year, the erring teacher was dismissed, along with a colleague who had publicly supported him.

Similar events took place in hundreds of other towns.

THE TEXAS BOOK REVIEW PROCEDURE

The longest running and best-known debates over textbook content took place before the Texas State Textbook Committee, which annually reviewed a large number of texts submitted by publishers and approved a list. Only books on the list could be purchased with state funds. For more than twenty years, Mel and Norma Gabler, a retired couple from Longview, Texas (with no formal education beyond high school but with sufficient financial backing to be able to employ full-time researchers) had been filing objections to the submitted texts. In their 1981 objections, they were reported to have challenged:

> open-ended questions that require students to draw their own conclusions; statements about religions other than Christianity; statements that they construe to reflect negatively on the free enterprise system (e.g. discussion of monopoly); statements that they construe to reflect positive aspects of socialist or communist countries (e.g. that the Soviet Union is the largest producer in the world of certain grains); any aspect of sex education other than the promotion of abstinence; statements which emphasize contributions made by blacks, Native American Indians, Mexican-Americans or feminists; statements which are sympathetic to American slaves or are unsympathetic to their masters; and statements in support of the theory of evolution, unless equal space is given to explain the theory of creation. (Weissmann 1982: 13)

The Gablers would like to 'get God back into the schools': 'The majority of Americans are Christians. We tell our children God is important, but how can they be expected to believe it if He is never acknowledged in the schools?' (in Weissmann 1982: 14). However, the Supreme Court has consistently and clearly ruled the teaching of religion in public schools to be unconstitutional. Hence the Gablers have been forced to move from promoting their own beliefs to attacking what they see as the alternative: secular humanism.

Thus many of their objections claim that material specified is an expression of the *religion* of secular humanism, and should be deleted as unconstitutional.

What has made the Gablers influential is the size of the Texas market relative to the rest of the country. No publishers were prepared to produce two copies of their texts, one for the South and one for the rest of the country. Rather than face long lists of objections from the Gablers and their supporters, major publishers changed their texts. Laidlaw, a subsidiary of Doubleday, purged the word evolution from its only high school biology text. As a Doubleday executive candidly explained: 'the reason for self-censorship is to avoid the publicity that would be involved in a controversy over a textbook. We'd like to sell thousands of copies' (in Bollier 1982: 194).

In 1983, People for the American Way took on the Gablers and won. PAW argued through the courts that the system was discriminatory and unbalanced in that the only permitted lay participation was objection. The Gablers (and others) could object to the submitted texts but ordinary people were not permitted to defend them. As a result of considerable agitation, the Governor took the opportunity of a demand for a general reform of the Texas school system to remove the Textbook Committee and to place the vetting of books in the hands of professional educators.

There is no doubt that liberal organizations such as PAW have exaggerated the impact of the Gablers, as have the Gablers themselves. In 1981 they claimed to have kept eleven social science texts from adoption. But only thirteen were being considered, only five could be adopted, and the Gablers had objected to twelve of the thirteen. Thus at least seven of the books which they found objectionable would have been rejected irrespective of their having objected to them!

Most of the texts which the Gablers have found offensive have been adopted. Even when the objectionable texts have been dropped, such decisions have often been made on grounds quite separate from the Gablers' objections. But, as Barbara Parker, the education project director of PAW, has consistently stressed, the most significant point of impact of protests such as those organized by the Gablers is not adoption but anticipatory self-censorship by publishers.[3]

[3] This point has been the main theme of PAW campaigning on censorship and

Although the influence of the Gablers has been reduced by a successful liberal campaign, the last ten years have seen a large number of small-scale local conservative victories. The cumulative impact has been for educationalists to play safe. Rather than face well-organized conservative opposition, school boards, heads, and teachers have chosen to refrain from the discussion of potentially contentious issues such as evolution and sex education. But even here, the success of the conservative movement should not be exaggerated. The 1983–4 People for the American Way report on censorship lists seventy-three incidents. In more than fifty of these, the conservatives failed in their objectives. The 1985–6 report shows a considerable increase in the number of incidents—almost double—but the compilers rightly explain this as a result of better reporting rather than a dramatic increase in censorship attempts. Of 124 incidents only thirty-four could be described as conservative wins and many of these were not concerned with the more contentious elements of the NCR platform but with matters such as 'obscene language' in school plays.

A large part of the early success was due to surprise. Liberals and professional educators either caved in under pressure from protesters or outrightly refused even to consider that some protests might be legitimate. As the large number of incidents listed in the PAW reports as having no clear outcome suggests, a more successful strategy for diffusing tension has gradually been evolved. Contentious books are withdrawn temporarily. Committees are appointed to hear detailed submissions. A degree of accommodation, such as placing certain books on some sort of restricted loan system, is offered. In most cases, the protests run out of steam and no further action is taken. Just as the conservatives had been quick to call on the services of professional protest organizations such as the Eagle Forum, liberals have created their own support organizations.

Recent surveys of attempts at censorship contain evidence for an often neglected observation: conservative Christians are not the only ones engaged in attempts to control the contents of books and courses. In the past, groups representing blacks and Jews have protested against racism and anti-Semitism. More recently, feminist

was argued for and instanced by Barbara Parker in a very large number of magazine articles, press releases, and reports on censorship between 1978 and 1984 (e.g. Parker 1979; 1983).

groups have become active in the field. NCR activists see themselves as doing nothing more than reasserting their right to be taken seriously as a legitimate minority group. It is no accident that much of the fundamentalists' anger is directed towards acts of affirmative action which accept as legitimate the rights of blacks, hispanics, homosexuals, or women as members of a group rather than as individuals. Although they have responded to the state's acceptance of group rights by asserting that only individuals have rights and that they are a 'moral majority' of such individuals, at the same time fundamentalists seem to have accepted their place as one more group claiming the attentions of the state.

Ralph Turner's famous comparative description of the American and British education system as offering 'contest' and 'sponsored' mobility is generally appropriate for conceptualizing the different ways in which the two societies organize many activities. Public administration in Britain is heavily centralized, élitist, and paternalistic. Although America is not the land of equal opportunity and open competition which its rhetoric would suggest, far more matters are arranged through combat. In the last resort the courts are required to adjudicate the combat and this has been the case with educational policy disputes. In 1982 the Supreme Court ruled on the general issue of parental right to limit the availability of certain books on grounds of ideological unsuitability. In the judgement, Justice Brennan said:

> local schools boards may not remove books from school library shelves simply because they dislike the ideas contained in those books and seek by their removal to prescribe what shall be orthodox in politics, nationalism, religion or other matters of opinion. (National Coalition Against Censorship 1982: 1)

Although there are sufficient loopholes in the judgement to permit conservative groups to continue their local campaigns, the fact that the matter came to a judgement of the Supreme Court shows clearly the limits which constrain regional actions, no matter how powerful their support, and demonstrates the Court's willingness to continue to judge regional particularisms (especially of a religious nature) to be unconstitutional.

CREATIONISM AND EQUAL TIME LAWS

As will be seen, two of the themes from the above discussion—the arbitration role of the courts and the fundamentalist acceptance of a part of the rhetoric of their opponents—are central to the analysis of recent conflict between creationists and evolutionists.

For a number of fairly obvious reasons, the belief that God created the world in the manner described in Genesis is of considerable importance to fundamentalists. The centre of the fundamentalist credo is the proposition that the Bible means what it says. What it says is that God made the world in six days. The age of the earth is also important. In order to maintain the general chronology which they derive from their Bible reading, fundamentalists need to suppose that the earth is about 10,000 years old. To accept the evolutionist view that the earth is over 4, 000 million years old would cause problems with the interpretation of prophecy, independent of it involving an admission that previous fundamentalist orthodoxy was mistaken. Additionally, fundamentalists believe that evolutionary theories, in denying that there is a radical division between man and other species, produce sin and moral degradation by encouraging people to act in a bestial manner.

In 1925, a biology teacher decided to challenge a Tennessee law which prohibited the teaching of Darwin's theory of evolution. John Scopes was arraigned for trial. The prosecution in the case was led by William Jennings Bryan, the great Democratic populist leader. With support from the American Civil Liberties Union, Scopes was defended by Clarence Darrow. Technically the case was won by Bryan, who died shortly after, but the argument—at least in the eyes of urban, educated, and cosmopolitan America—was won by the evolutionists. From our vantage point, what was important about the case was not the decision. With a fundamentalist jury, three members of which testified that they read nothing but the Bible, the verdict was a foregone conclusion. What was important was the nature of the prosecution case. Bryan attacked evolution on two grounds. He asserted the supreme authority of the Bible and, as one would have expected with his populist background, he declared his faith in the superior judgement of the common man with common sense. Bryan was not concerned with what scientists thought. If science said one thing and the Bible said the other, he had no doubt which was correct: 'It is better to trust in the Rock of Ages, than to know the age of rocks; it is better for one to know that he is close to the Heavenly Father, than to know how far the stars in the heaven are apart' (in Marsden 1980: 212). He was also unimpressed by the status of the professional scientists and the liberal theologians produced as witnesses for the defence. Just as Protestantism rejects the priesthood, so fundamentalism distrusts the college-educated minister, the man with too much learning for his own good. The Bible was the word of God and he had written in such a way that ordinary people could understand. In a nutshell, Bryan and his supporters were openly anti-intellectual and anti-scientific.

By the late 1960s, creationists had radically changed the grounds for their attack on evolutionism. The Genesis account was now presented as 'creation science'. The Creation Research Society was founded with its membership restricted to the holders of at least a second degree in some science. Fundamentalists began to exploit disagreements between evolutionists to argue that evolution was at best an interesting hypothesis. If evolutionists presented their theories as if they were established 'laws', it demonstrated that they believed in evolution for non-scientific reasons. Thus evolution was, in fact, a religion.

With less conviction and far less detail than the critique of evolutionism, the creation account was presented as a plausible scientific hypothesis which fitted with the known facts at least as well as did evolution. Instead of quoting the Bible, creation science publications combined critical analysis of evolutionist writings with the sympathetic presentation of those pieces of the factual record which could be read as supporting special creation. For decades there has been argument about the correct interpretation of tracks in the seasonally dry river beds at Paluxy Creek, Glen Rose, Texas. Some of the tracks are clearly dinosaur prints but alongside those are other tracks which do not show the clear impression of three toes. Creationists argue that these are human prints which prove that dinosaurs and humans roamed the earth together. Evolutionists insist that they are simple dinosaur prints with the toe marks eroded or filled in.

Crucial to the new-found confidence of creationists was the gradual change in the claims made for the products of scientific thinking and research. At the start of this century, scientists tended rather readily to suppose that they had 'proven' things to be the case and many ordinary people supposed that they had. The scaling down of claims, given its most articulate form in Karl Popper's refutationalist philosophy of science, gave vital leverage to the creationists. When cautious scientists of a Popperian bent announced that they could not conclusively *prove* anything, but could only refute previous mistaken views and thus inch forward, creationists repeated such statements but did so without making clear the epistemological and methodological reasoning which informed the rather special use of 'prove' in such tentative claims. To the lay listener, 'cannot prove (in the hard sense) *X* to be the case' can sound very like '*X* is not the case' or 'We do not know what is the case'. Long after natural scientists had abandoned a Baconian vision, creationists continued to see the world in crude positivistic terms and used the reluctance of scientists to make grand claims as evidence of their inability to support their modest ones.

It is of no great consequence, but there are interesting inconsistencies in the creationists' use of Popper. They are happy to misunderstand evolutionists when such scientists, on Popperian grounds, refuse to claim to have proved evolution. And they are fond of citing Popper's criticism of Darwinism that it was an historicist and irrefutable metaphysic rather than a scientific theory (Popper 1978: 167–80). However, they ignore Popper's later clarifications and are reluctant to address the question whether creation science would be a science if it were judged in Popperian terms.

In making the following point I do not wish to suggest that creationists have cynically adjusted their rhetoric. There can be no doubt that they believe the Genesis account to be correct and hence 'scientific', but a further good reason for the shift from creation*ism* to creation *science* was the clear message from the Supreme Court that the constitutional requirement for a separation of church and state made the teaching of Genesis 1–12 in biology classes illegal because it constituted state support for religion. Clearly, if the Genesis account was to be restored to public schools, it would have to be presented with a novel justification; it should be taught primarily because it is scientifically respectable and not because it is the Word of God.

In keeping with the NCR's perception of itself as a legitimate minority, the creation science case was also presented as a matter of fairness. The ideas of the evolutionists, which were no better, were taught in schools. Fairness required either that evolution be banned from the classroom or that creation science be given 'equal time'.

Some idea of the extent to which such an approach would be given a sympathetic reception is conveyed in one survey of white middle-class Texans. Although 62 per cent wanted evolution to be taught in public schools, 73 per cent said that creationism should also be taught (Stacey *et al.* 1982).

The claim for fair treatment has been surprisingly successful. In most other spheres of life, lay people defer to the scientific experts. There is a massive consensus among scientists that some sort of non-supernatural account of the origins of the earth is considerably more plausible than the idea of divine creation. Yet many people in the most technologically advanced and wealthiest nation in the world seriously consider the claim that fairness requires that a theory supported by all the scientists with any professional standing in the matter should be countered by an alternative supported only by people who have no claim to expertise. And the claim has not only been taken seriously but it has been accepted by the legislatures of a number of states.

THE ARKANSAS EQUAL TIME BILL

The career of the Arkansas bill can be used to highlight a number of themes. Like the Kanawha County textbook controversy, the Arkansas law was not the consequence of any carefully constructed legislative programme promoted by a national NCR organization. Paul Ellwanger, an X-ray technician from South Carolina, who was neither a lawyer nor a scientist, wrote a rough draft of a legal measure requiring that any teaching of evolutionary theories for the origins of species be balanced by equal time given to creation science. Refined by some creationist friends, the draft was circulated to the membership of a small organization called Citizens for Fairness in Education which Ellwanger ran from his home. One of the members who lived in Arkansas arranged for the draft to be introduced in the state legislature where, at the end of a very long sitting, it passed both houses with barely any comment. The Democratic Governor of the state, himself a fundamentalist, happily signed the bill into law as Act 590. Later, when the contents of the Act became known, he said that he had not read it carefully before signing.

Gene Lyons is a skilled, if critical, journalistic commentator on fundamentalism and a resident of Little Rock, Arkansas. He is probably right to stress the lack of zeal behind the passing of Act 590:

> Arkansans in general are probably no more ignorant than the American public at large, but all the ignoramuses do agree. Political tradition here pardons a legislator who votes on symbolic issues to soothe the prejudices of the mouth-breathing element in the dirt-road churches. Arkansas is more than 90 percent Protestant, the hard-shell sects predominate, and ambitious youth yearn to be television evangelists as others wish to emulate Reggie Jackson or Donny Osmond. No sense, runs the usual logic, in stirring people up; the federal courts can take care of it. Then everybody can whoop it up in the next campaign about meddlin' judges thwarting the will of the people, can get reelected, and can continue to work on the truly important business of democracy, like exempting farm equipment from the sales tax or allowing the poultry industry to load as many chickens as can be jammed into a semi-trailer regardless of highway weight limits. Indeed it appears that many of the legislators mistook the creationist bill for yet another in the series of harmless resolutions in praise of Christianity that they customarily endorse. (1982: 39)

Indeed the federal courts did take care of it and the conservatives did complain about 'meddlin' judges'.

The trial ruthlessly exposed the weaknesses in the fundamentalist position. In defence of its claim that the law was unconstitutional because it demanded state support for a religion, the American Civil Liberties Union was able to call a stream of distinguished and articulate scientists. Although it was clear from their testimony that there were major differences between different schools within the evolutionary camp, and that there were large gaps in the evidence for the theory, it was also abundantly clear the doubts were about details and not about the broad framework. That much was only to be expected. Where the creationists (and the unfortunate state Attorney General who had the job of defending Act 590) ran into unexpected problems was with the poverty of their own case. They had to establish that there was a body of knowledge—creation science—which was plausible *independent* of a particular religious world-view. Unfortunately two of the more articulate advocates of creation science—Henry Morris and Duane Gish of the Institute for Creation Research—were barred from arguing this case by their own published writings. Morris had written 'Creation cannot be proved [because] Creation is inaccessible to the scientific method' and Gish said:

We do not know how the Creator created, what processes He used, for *He used processes which are not now operating anywhere in the natural universe*. This is why we refer to creation as special creation. We cannot discover by scientific investigations anything about the creative processes used by the Creator. (in Lyons 1982: 75)

While Gish's position is consistent (he argues that both evolutionism and creationism are religions), it hardly does much to establish what Act 590 called 'the scientific evidences for creation'. By all accounts (including those of some fundamentalists) the witnesses who were called to argue for the existence of creation science were deplorable. The Act had been framed in such a way as to saddle the Attorney General with the defence of the 'two model approach'. This argued that, as there were only two possibilities—atheistic evolutionism and theistic creationism—anything which did not support evolutionism must be support for the creation account. At one point in his cross-examination of Professor Francisco Ayala, a world-renowned geneticist from the University of California at Davis, the Attorney General mentioned the 'two models' and Ayala replied:

'Son, in science it is impossible ever to say there are only two models or theories. Everything is always open; new ideas, new vistas, new perspectives, new forms of inquiry, are always appearing. No one of these is closed if it makes sense—and never are there only two possibilities. (in Gilkey 1985: 141)

When the Attorney General (a Mr Williams) suggested that criticisms of evolution were positive arguments for creation science, Ayala replied:

My dear young man, negative criticisms of evolutionary theory, even if they carried some weight, are utterly irrelevant to the question of the validity or legitimacy of creation science. Surely you realize that *not* being Mr Williams in no way entails *being* Mr Ayala. (in Gilkey 1985: 141)

This tack was soon abandoned and the state's attorneys were left trying to draw from their witnesses any good reason, other than the fact that the Bible said so, for believing in special creation. One witness insisted that the scientific community was too prejudiced against creationists to publish their works. When pressed by the judge, he was unable (as was anybody else) to give any evidence of a creationist paper which had ever been submitted to a refereed scientific journal. Finally, having made so much play of the disagreements between evolutionists, the creationists had to listen to one of their star witnesses, mathematics lecturer N. K. Wickramasinghe of the University of Wales, describe most creationist ideas as 'claptrap'.

In his judgement, William Overton concluded that there was no creation science independent of religiously inspired beliefs in the Genesis account of the creation. Hence Act 590 was unconstitutional in that it required the public schools to teach a particular religion. The Arkansas Attorney General decided not to appeal against Overton's decision.

An equal time bill similar to the Arkansas law was passed by the state of Louisiana. It was similarly challenged and found unconstitutional by the courts but the Louisiana Attorney General felt sufficiently committed to the Act to appeal the decisions of the lower court. In June 1987, the Supreme Court by the considerable margin of 7 to 2 upheld the decision of the district court, and judged that, despite statements of the Louisiana legislature to the contrary:

> . . . the purpose of the Creationism Act was to restructure the science curriculum to conform with a particular religious viewpoint . . . [*and*] because the primary purpose of the Creationism Act is to endorse a particular religious doctrine, the Act furthers religion in violation of the Establishment Clause. (Supreme Court of the United States, 85–1513, 19 June 1987: 13–14)

THE MORALS OF THE EQUAL TIME CASES

There is no doubt that the creationists' appeal to a sense of fair play in framing their demand as the right to equal access or equal time was successful. Perhaps Act 590 was helped into law by a tired and overworked legislature's unwillingness to look too closely at an apparently uncontroversial endorsement of God, but similar bills were proposed in fifteen other states. President Reagan himself endorsed the principle of equal time and 76 per cent of respondents in an Associated Press–NBC poll agreed that both theories should be taught in schools (*Guardian* 3 December 1981).

The problem for the equal time argument is that, outside very small and culturally homogeneous areas, there will always be people who will object to creationism. As they were in Arkansas and Louisiana, equal time laws will be considered by the courts. In the full glare of public attention, creationists will have to put detailed flesh on the rhetorical bones of creation science. At this point, the accommodation which the creationists have made with secular models of knowledge—the claim to be scientific—will come under relentless scrutiny. With the failure of the 'two models' approach to explanations of the origin of species, it will not be enough to show inconsistencies and disagreement within the evolutionist camp. Creation science will have to demonstrate itself plausible to any reasonable non-believer, and to date it has been unable to do this.

Furthermore, it is clear that the early successes of the creation science approach depended a lot on surprise. Evolutionists were neither practised at, nor prepared for, a public defence of their views. Having supposed that creationism had been irrevocably discredited in 1925, they took some time to see the need for articulate presentations of their ideas. However, by the middle of 1982, such well-known figures as Stephen Jay Gould were touring the country, appearing on television and radio shows, writing 'op-ed' pieces for local newspapers, and generally engaging in a well-organized public relations exercise to combat creationism.

The Arkansas and Louisiana cases also show the limits to the appeals which can be made to democracy. That a lot of people believe in special creation is enough to cause some publishers to be cautious about advertising Darwinian evolution in their biology textbooks. It is also enough to introduce equal time laws. But for all that American judges are more electorally vulnerable than their British counterparts, their own shared beliefs and values are relevant to their decisions. Although federal judges are political appointees they must be minimally qualified. All senior federal judges will have studied at one of the major law schools and most will have practised for many years in large cities. They are cultured university-educated members of the upper middle class. Although some senior justices are extremely conservative in their politics, none (to my knowledge) sympathizes with the view that creationism is anything other than a religious belief. The 7 to 2 Court verdict against the Louisiana bill suggests the size of the hurdle. Even if a significant number of judges and justices did privately share creationist beliefs, they would need far more convincing performances from creation scientists before they could do anything other than rule equal time bills unconstitutional.

THE CONSERVATIVE APPEAL TO THE CONSTITUTION

With attempts to present their own beliefs and express their own values in schools apparently blocked by the Supreme Court, fundamentalists are trying to find another route to the same destination by more fully exploiting a sense of fair play. In 1983, Judge W. Brevard Hand of Mobile, Alabama, challenged several Supreme Court judgements with a ruling which permitted school prayer by arguing that the First Amendment did not apply to such cases. His argument was based on a return to the earliest interpretations of the constitutional requirement for the separation of church and state: that it applied only to the federal 'state' and not to the individual states. He was overruled on appeal. His response was to give the plaintiffs guidance on how they might reframe their case. With assistance from televangelist Pat Robertson's National Legal Foundation, 624 fundamentalists filed a suit against the Alabama State Board of Education, charging that the Board had violated their constitutional rights by teaching the *religion* of secular humanism.

The logic of this is the same as that of the idea that there are only two models of the origins of the species; the world of knowledge divides into two competing camps. There is true knowledge—theism—and there is anything else. Anything which attempts to explain or even just to describe parts of the world without reference to God must be secular humanism. The state, by excluding conservative Protestant beliefs from schools, is deliberately promoting a competing alternative. Hence fundamentalists can present themselves as members of a minority whose constitutional right to religious freedom is being denied by the operations of the state.

Given Judge Hand's previous judgements there was little doubt that he would find for the fundamentalists. But he was overruled by the unanimous verdict of three Appeal Court judges who refused the 'two theories' approach and considered on its own merits the claim that the books in question taught secular humanism. They concluded that they did not. As with the creation science argument, fundamentalists are pursuing the logic of the liberal endorsement of minority rights to establish themselves as a legitimate minority. But as in the creation science cases, there is a high probability that this strategy will prove self-defeating. An initial symmetry between theism and secular humanism is easy to suggest with well-arranged prose, but it is harder to establish when such prose is rigorously challenged.

Fundamentalists often assert that the Supreme Court has already determined that secular humanism is a religion (see above, p. 78). Detailed examination of the nature of the secular humanism which presently offends fundamentalists demonstrates that, far from having any coherent dogmas, it is simply an agglomeration of anything which fundamentalists do not like which is not already some other religion. That is, it can only be defined by reference to what it is not. It is illustrative of the difficulty of establishing the identity of secular humanism that the plaintiffs in the Alabama case had to call on a sociologist to testify that secular humanism is the 'functional equivalent' of a religion. While giving with one hand, James D. Hunter rather took away with the other by adding that 'vegetarianism, socialism, environmentalism and bureaucracy' might also be seen as functional equivalents for religion (*Time* 27 October 1986).

There may be something essentially correct about the fundamentalist argument that to teach about things on which they have strong views, without endorsing those views, is, *de facto*, to deny the relevance, if not the validity, of fundamentalist beliefs. However, there is no possibility that fundamentalists can turn that commonsensically sound notion into a justification for treating all non-fundamentalism as a competing religion which should be banned from schools. Unfortunately I cannot remember the cartoonist but I recall a marvellous drawing of a fundamentalist child rather mischievously asking his mother: 'My mathematics homework doesn't mention Jesus. Does that mean I don't have to do it?' As the higher courts are bound to conclude, were the 'that which isn't our religion is a competing religion' argument to be pursued to its logical conclusion the state would be unable to promote any beliefs about anything. There would be nothing left which the public schools could teach. The issue seems to have been fairly well settled by a recent judgement of the Sixth Circuit Appeals Court on a Hawkins County, Tennessee, case. In overturning the decision of a lower court, the three Appelate judges unanimously stated that requiring children to read books which did not endorse their beliefs, or even which challenged their beliefs was not an infringement of their constitutional rights. They ruled that the earlier decision had failed to distinguish between simply reading or talking about other beliefs and being compelled to adopt them. As one of the judges argued: 'There was no evidence that the conduct required of the students was forbidden by their religion' (*Time* 7 September 1987).

CONCLUSION

To summarize these points in terms of the structural differences between Britain and America outlined in Chapter 3, the open nature of the American legislative and electoral systems allows well-organized interest groups (and, in the Arkansas case, not particularly well-organized interest groups!) to introduce laws which will protect or promote their particular culture, and to make their agenda an election issue. It is easier for American conservative Protestants to place their concerns on the public agenda than it is for their British counterparts. In particular they are able to utilize the rhetoric of minority rights and fair play initially to predispose some members of legislatures and the judiciary to give them a hearing. Clearly the arguments have not yet finished. There is every reason to suppose that conservative Protestants will continue to raise legal actions, to press candidates to embrace their positions, and to introduce their bills to state legislatures. In ending this chapter with discussions of cases where the NCR has failed to achieve its goals, we should not forget Parker's point about textbook censorship, which can be generalized to embrace the whole NCR agenda. Even when cases are lost, the very fact of their having been raised has the subtle effect (which, unfortunately for social scientists, cannot be measured) of shifting the centre of public discourse as publishers, teachers, school administrators, and politicians play safe.

However, as will be elaborated in the final chapter, there are good reasons to suppose that conservative Protestants will continue to be defeated by the modern secular world. The creation science debates suggest that adopting some of rhetoric of the modern pluralistic liberal society is self-defeating because it is asking to be judged by the rules of a game in which religious orthodoxy is an irrelevance.

6

Lamentations: Sources of Weakness

THE difficulties of making a precise diagnosis of the health of the new Christian right in 1988 are such that this and the remaining chapters will occasionally display a slight awkwardness of tense. The general purpose of this discussion is to provide an alternative to the two most common responses to the NCR. On the one hand, its most ardent supporters and critics exaggerate its significance and list its strengths in such a way as to make it difficult to see why it has not already swept all before it. On the other, social scientists such as Tamney and Johnson suggest that the movement is little more than a social myth born of overreaction to a successful campaign by a small cadre of self-appointed leaders. The previous chapters have, I hope, demonstrated that a movement which can usefully be called 'the new Christian right' exists. This chapter will point to previously overlooked sources of fragility in the movement and attempt a more accurate assessment of it. By the time many readers come to this text, the NCR will, in its present form, be dead. However, given that efforts to mobilize conservative Protestants will continue in some form or other, the present tense will be used to describe those internal weaknesses which from some future vantage will be perceived as part of the explanation of the present movement's demise. This chapter can be read as a doctor's report on a sick patient; with the tenses changed from the present to the past, it may in the not too distant future be read as an autopsy.

COMMITMENT AND COMPROMISE

If the NCR was to have any lasting impact, it needed to add conservative Catholics, Mormons, and Jews to its starting support-base of conservative Protestants. To date the movement remains largely Protestant. In order to cultivate and sustain links with other conservative elements of what Falwell calls 'our shared Judaeo-Christian' heritage, the NCR must encourage a number of attitudes amongst its members and I will suggest that such attitudes produce problems of sustaining commitment; that is, the conditions for expanding the movement are also conditions which weaken its core.

The NCR needs to *compartmentalize*, to keep separate religion and the moral crusades which the religion has produced. Although fundamentalists are opposed to abortion on religious grounds, they must be prepared to work with those who share their dislike for abortion, but who do not share their religion. While this sort of compartmentalization is characteristic of modern society, it goes against the grain of fundamentalism, which refuses to accept that religion can be kept in a box marked 'family' and 'Sunday'. While some of the leaders of the NCR seem able to compartmentalize, other fundamentalists (such as those associated with Bob Jones University) have been highly critical of what they see as preaching morality without preaching the gospel.

Successful compartmentalization is a threat to the commitment of grass-roots supporters because they have been recruited to the movement through their religious ideology which, in order to make the movement effective, they are now supposed to lay to one side. At the same time, the failure of the fundamentalists to compartmentalize successfully is an obstacle to the participation of conservative Catholics, Jews, Mormons, and others because, on the occasions when the compartments break down, they are reminded of how fundamentalists really think of them. For example, doubt is cast on Falwell's performance as leader of an organization which claims to represent 'our shared Judaeo-Christian tradition' when it is made known that in his Moral Majority manifesto he said 'If a person is not a Christian, he is inherently a failure' (1979: 62). He was also widely reported as telling an audience of Virginian fundamentalists: 'A few of you here today don't like Jews. And I know why. He [*sic*] can make more money accidentally than you can on purpose' (Conway and Siegelman 1982: 168). The chairman of the New York state Moral Majority was similarly anti-Semitic: 'I love Jewish people deeply. God has given them talents He has not given others. They are His chosen people. Jews have a God-given ability to make money, almost a supernatural ability to make money . . . they control the media, they control this city . . .' (in Bollier 1982: 88). The theologically orthodox fundamentalist position was given by Bailey Smith, then President of the Southern Baptist Convention: 'It's interesting at great political rallies how you have a Protestant to pray, a Catholic to pray and then you have a Jew to pray. With all due respect to these dear people, my friends, God Almighty does not hear the prayer of a Jew' (in Bollier 1982: 89). Jimmy Swaggart, a popular young televangelist in the loud, pulpit-thumping mould, supports the NCR while continuing to denounce Catholicism as a 'false cult'. He also appears to believe that God permitted the Holocaust as a punishment for the Jews' rejection of Jesus Christ (PAW 1986: 3).

The most pressing problem for the NCR is to override such particularism and create a genuine alliance with conservative Catholics, Mormons, and Jews. However, the compartmentalization of religion and politics which such an alliance requires is precisely the social psychology which fundamentalism exists to oppose. Doing politics in a modern Western democracy requires the pragmatic separation of religion and politics but it is the very religious beliefs which caused the initial support for the NCR and which still serve as the main source of motivation for involvement. Successful compartmentalization threatens to undermine fundamentalists' commitment to political activism while the failure to compartmentalize threatens the movement's ability genuinely to broaden its support base.

COMMITMENT AND ACCOMMODATION

The problems of working with secular conservatives can be described under the headings of pragmatism and accommodation. The NCR leadership has been embarrassed a number of times by its more enthusiastic supporters. Falwell disowned the Baltimore, Maryland, chapter of the Moral Majority for picketing a bakery which allegedly produced 'sexually explicit' gingerbread men. A chairman of the Moral Majority in northern California had to be reprimanded for suggesting the death penalty for homosexuals. Accommodation to the prevailing norms will increase secular conservative support but it will also cause the grass roots to doubt the movement's orthodoxy. As early as 1982, significant numbers were leaving and a California chapter of the Moral Majority seceded because the national organization was becoming too 'liberal' (Lienesch 1982: 418).

In his political associations, Falwell has shown signs of accommodation. In 1979 Falwell and others were strongly hostile to the presidential ambitions of George Bush, who was widely seen as a representative of east coast establishment Republicanism. But in 1986, while other conservatives were promoting the cause of Jack Kemp—the favourite in a 1986 *Conservative Digest* straw poll—as Reagan's successor, Falwell publically endorsed Bush. Once fellow television evangelist Pat Robertson made his interest known, Falwell qualified his support for Bush. A conservative and sympathetic biographer reported, without denying it, the feeling that Falwell was moderating:

> And some said that [*the endorsement of Bush*] reflected Falwell's desire to become part of the respectable mainstream in American political life. Many of Jerry Falwell's people are tired of being treated as outcasts; they want to be accepted by the rest of the country, and if a moderated political approach to personalities can help achieve that, some are willing to live with it. (D'Souza 1986: 12)

A degree of moderation can also be discerned in the career of Pat Robertson who founded the Christian Broadcasting Network (CBN) and who hosts the popular fundamentalist chat show, the *700 Club*. Robertson's relationship with the NCR has always been ambiguous. Initially involved with the symbolically important Washington For Jesus rally in 1980, he later announced that God wanted him to 'back away' from politics. In turning down my request for an interview in 1983, Robertson asserted that neither he nor CBN were 'in politics'. However, he continued to accept invitations to be involved in such NCR organizations as Tim La Haye's Council for National Policy and he created a number of his own organizations. One was a legal foundation which provides expertise and resources for fundamentalists promoting their issues in the courts.

When in 1986 Robertson began seriously to pursue the Republican presidential nomination, it was noticeable that his political managers engaged in biographical reconstruction. Consider the following biography written by Howard Phillips as an introduction to an interview with Robertson in *Conservative Digest* (in which I have inserted letters to identify the significant points).

> American entrepreneur [*a*] M. G. Pat Robertson is the son of a distinguished US Senator [*b*] and descendant [*c*] of two Presidents of the United States. Graduated *Phi Beta Kappa* and *magna cum laude* from Washington and Lee [*d*], he fought as a Marine lieutenant in Korea [*e*],

earned a jurisdoctorate from the Yale Law School [f], and then a master of divinity degree from the New York Theological Seminary [g]. A former Golden Gloves middleweight contender [h], this son of Tidewater Virginia then had the heart and commitment to minister [i] in the ghetto of Brooklyn's Bedford–Stuyvesant district [j]. (Phillips 1986: 85)

Of the ten qualities or characteristics which Phillips wished the reader to notice, only two concerned Robertson's religion. One of these was presented as an item in a list of educational credentials and the second—'heart and commitment to minister'— was offered not as evidence of piety but as a story about general character.

In a roundtable discussion of the Reagan legacy (Duke 1986: 77), a number of political commentators drew attention to Robertson's image-making:

Johnson: I noticed that Robertson, for instance, was quick to point out after a recent profile of him in the *Wall Street Journal* that, yes, he's an evangelist all right, but he's also a college graduate, a son of a senator, Phi Beta Kappa.
Duke: He went to Yale.
McDowell: And the London School of Economics.
Johnson: In other words, he's not just somebody who found Jesus.
Duke: Are you saying that he's beginning to downplay his religious base?
Johnson: Sure, yeah. He obviously senses that there's a problem if you're seen *only* as an evangelist when you're running for President.

In meetings on university campuses in spring 1986, Robertson chose to speak on constitutional rather than specifically NCR issues. Instead of attacking the particular decisions of the Supreme Court to which the NCR object, he criticized the power of the Court. When speaking at the University of Virginia, it was not until a questioner from the floor raised the matter of his faith that Robertson gave the sort of testimony which previously would have been the starting-point of any public discourse. Equally revealing is an exchange in an interview with a *Conservative Digest* staffer (McGuigan 1986c: 34). After a discussion of his own religious views, Robertson was asked the following question, a clear invitation to criticise the mainstream churches:

McGuigan: While we are on this topic, what about the 'mainline' Protestant churches? Or have they become so secularized that they do not matter anymore?

Robertson: I don't think the positive changes that have taken place in this country in recent years have been restricted to any select group. We welcome support from every American of good character.

Such a polite, guarded, and circumspect response represents a considerable departure from classic fundamentalism which requires, almost as proof of orthodoxy, that every opportunity be taken to criticize the apostate denominations.

A political scientist said:

> Falwell . . . has moved toward the political center, following a familiar pattern among leaders of extreme factions who, once they gain fame, recognize that responsible use of power in a pluralist democracy requires moderation. To put it more critically, they begin to crave acceptance within the moderate establishment. (Reichley 1986: 28–9)

While the general description of what was happening is correct, there is an alternative explanation which does not denigrate the motives of Falwell and his colleagues. Firstly, while they may personally crave acceptance within the establishment (they are certainly fond of photographs of themselves with the President and senior officials), they also crave acceptance for their principles and issues. Secondly, they may well be accommodating, not because they want to be popular, but because they believe pragmatic compromise is essential for political success.

There were particularly bitter internal arguments in the NCR about the best way to proceed with the anti-abortion campaign. Since the middle of the 1970s, pro-life organizations had been trying to pass measures to outlaw abortion completely and to establish conception as the legal 'start of life'. By the end of 1982, Falwell was shifting from what Ron Godwin, one of the key administrative figures in the Moral Majority, called the idealist position. As an alternative he promoted the more pragmatic view that effort should be concentrated on measures to restrict the availability of abortion. An indication of the seriousness of the opposition to such a shift is the amount of effort which the Moral Majority put into defending the change of direction. One issue of *Moral Majority Report* devoted its centre spread to a long defence by Godwin of Falwell's pragmatism. It is clear from the content of the defence that Falwell's personal values were being called into question:

> Purists may accuse Falwell of bad judgement and say that his pragmatism won't work, but they cannot pretend that he cares less for innocent life than do they or that he has abandoned his convictions simply because he assesses the mood of Congress differently than they do. To make such a judgement of his motives is indefensible and reflective of a pattern of behavior that is denigrating to the public image of the movement. (Godwin 1983: 4)

Without much reading between those lines, one can see evidence of the classic conflict between the enthusiasts and the pragmatists; the enthusiasts claiming that Falwell has 'sold out' and the pragmatists insisting that some limited success would be better than noble failure. Additionally Godwin deployed the standard plea for solidarity: the movement's goals will be made less rather than more easy to attain if the enthusiasts insist on placing principle (however noble) before pragmatic adjustment to the realities of the political situation in which the movement must operate.

There is in the sociology of social movements literature an excellent tradition of work on 'goal transformation' which begins with Weber's writings on the rise of bureaucracy and Michels's claims that even the most democratic voluntary associations move towards oligarchy. A number of studies have argued for a common career for social movement organizations.[1] As the movement grows beyond the size where business can be conducted by members in personal interaction, it comes to rely more and more on a cadre of professional officials who are more concerned with maintaining the organization than with attaining its goals. The initially radical goals are transformed into something more moderate and the satisfactions of being a member come to be more important than the sense of participating in a crusade which is close to success. For a range of reasons which need not be considered here, the NCR does not fit the standard goal transformations model particularly well. The initial goals of the movement were always so broad that quite radical changes of direction could be described as a change of emphasis rather than a change in goals. Furthermore, the NCR is not a single social movement organization. It is a general movement consisting of a variety of particular movements, each of which consists of a variety of organizations.

Although Godwin described the change in abortion campaign

[1] For elements of the goal transformation model, see Weber (1947/1964), Michels (1959), Zald and Ash (1966), and Wilson (1970).

tactics as being one of a move from idealism to pragmatism, it would be more accurate to say that most elements of the NCR leadership were always pragmatic. They always knew that political activity would involve getting their hands dirty. The shift on abortion (and on other issues) could best be described as a case of having to be more accommodating than had initially been anticipated.

But, and this is the main point, while the leadership was always in the compromise business, the membership was not. People are recruited to social movements by appeals to their idealism and their enthusiasm. This is especially the case with the NCR where the supernatural base of the values which produced the recruitment gives rise to zealotry. Compromise and accommodation may be necessary for any sort of success on the NCR agenda but it is not what people expected when they were attracted to the movement.

COMMITMENT AND EXPECTATIONS

If religion and politics require different criteria for judging the acceptability of others, and a difference in willingness to mute distinctive features of one's ideology in order to succeed, they are also distinguished by different notions about the certainty of knowledge and about time-scales.

Generally speaking, religion is about certainty. It is about knowing how to attain salvation and, until the relatively recent rise of liberal versions of Christianity and Islam (and with the possible exception of the more philosophical forms of Buddhism), it has been about *dogma*. Fundamentalists are perfectly traditional in their belief that they have perfect assurance of being saved. In contrast, politics are about fudging and compromise. Successful politicians are not dogmatic, even if they may be willing temporarily to appear committed to a dogma which looks like a vote winner. The balances and checks of the American political system were designed for inertia and most politicians are happiest when doing nothing because to do something will offend some group of voters or other. Without a strong party system to blame for their actions, most American politicians are loath to be too closely identified with particular ideologies and positions. Richard Neely, who was a member of the West Virginia state legislature before being elected to the state's Appeals Court, quite correctly points out that American legislatures are designed to allow the majority of introduced bills to die of neglect. The complex system of committees and scrutiny procedures allows politicians to be 'in favour' of a great deal of legislation without ever being required to vote on it, or even worse, actually pass it (1981: 23–78).

Those whose careers survive the zealotry common to the young learn that, if they do have to do something, they have to *deal.* Politics in a system as complex as American democracy is very rarely about winning or losing. It is about winning more than one loses or at least being able to tell the voters back home that one won as much as was possible.

> Regardless of what else one says about elected politicians, they are experienced. They may be experienced in the same way Moll Flanders was experienced, but they are where they are because they have learnt to keep most of their constituents happy most of the time. They understand the concept of compromise, the middle-way of the half-loaf, and have a healthy sense of the mix of incompatible, selfish, interests of which society is composed. (Neely 1981: 106)

There was always considerable potential for tension between the grass-roots fundamentalists with their dogmatism and the more pragmatic politically minded activists (especially those who worked in the lobbies in Washington, many of whom had the revealing habit of combining affection and disparagement in their references to 'our more hot-headed brothers'). The longer the movement existed after its initial much-heralded successes, the greater the tension, and the greater the frustration and disappointment, even with those politicians who are supposed to be on the side of the angels.

It is unfortunate that there is no large-scale quantitative data on the reactions of grass-roots conservatives to the Reagan administration. There is evidence that, after 1986, both Pat Robertson and Jerry Falwell had to cut back programmes because of a shortfall in funding, and Viguerie has been forced to shift the balance of his work from ideological to commercial direct-mail campaigns. Merle Black, a political scientist at the University of North Carolina commented: 'There may be a limit to the amount of money people are willing to give without seeing results. Reagan has been in office six years now and they still haven't seen much success with their agenda' (*Newsweek* 14 April 1986). It might be argued that the direct mailers simply overreached themselves but it cannot be without significance that even some of the experienced lobbyists, those with a good sense of how difficult it is to get things done, were expressing dissatisfaction after only a few years of Reagan. Viguerie, returning to his favourite theme, told the *Wall Street Journal* (14 March 1984) that, if the Republicans were 'only going to use conservatives as cannon fodder to raise money, to work precincts but not to govern, at some point there's going to be a new demand for a new party'. In a similar vein, Weyrich said 'we kid ourselves if we think Ronald Reagan's re-election will bring major gains for the conservative movement' (*Group Research Report* March 1984).

All social movements have problems fulfilling the expectations of their supporters, but in the case of the NCR there is an additional feature likely to amplify the sense of frustration over the length of time it takes to produce major social change. Fundamentalists were mobilized by a combination of dramatic threats and promises: unless America is turned around now, it will suffer God's wrath. If Christians get involved and 'we can get God back into the classrooms', as one NCR activist put the school prayer issue, 'then we can see an end to drug-taking and violence and teenage pregnancies and all that illiteracy. We will see an end to crime and my wife will be able to go downtown shopping in the evenings again.' Religious believers can think in these terms because they believe in an active creator God who can, if he wishes, change people and worlds. Experienced politicians know that it not only takes a long time to produce any significant change in public policy, but it takes even longer for policy change to produce effects, and often the effects are not the ones which were anticipated. Although it is a gross over-simplification, the point can briefly be made by contrasting the spheres of religion and politics as being characterized respectively by optimism and realism. The failure of many grass-roots Moral Majoritarians to appreciate this has led to some becoming dissatisfied with seven years of Reagan, who after all was 'their President', and nine years of active campaigning which have produced no obvious change. As early as 1982, Gary Jarmin, the legislative director of Christian Voice, was pointing to the potential for disillusionment, while adding an interesting qualification:

> [*Conservative Christians*] are involved now because of their interest in key issues such as anti-ERA, anti-gay rights, anti-pornography, and pro-family.

But their degree of involvement in the future will depend on whether the President and members of the Republican Party pay attention to these moral issues—regardless of whether they win the legislative battles. (in Latus 1984: 258)

What is interesting in Jarmin's observation is the possibility that reactions may have less to do with actual results than with perceptions of the sincerity of conservative politicians. It is certainly my impression that some grass-roots new Christian rightists can temper their frustration at lack of legislative progress if they can continue to believe that the 'good guys' are still good. This shows itself in a tendency among conservative Protestant Reagan supporters to accept at face value his commitment to their positions and to explain his failure to deliver as a consequence of the Democrats' control of Congress. There is some justification for this. *Congressional Quarterly* calculates a presidential success rating by collating the fates of measures to which the President has given clear support. In his first year Reagan had a success rate of 82 per cent. In 1982 this fell to 72 In 1983, it was just 67 per cent. In contrast, Jimmy Carter, in his final year when he was widely perceived as a weak President embroiled in the Iranian hostage crisis, managed to win three-quarters of his battles with Congress.

However, it has not escaped the notice of some conservatives that the President has been reluctant to use his influence to press legislation on socio-moral issues. The politically astute must have appreciated that the Laxalt–Hansen Family Protection Bill (introduced in 1983) was bound to fall because it was too broad. With the purpose to:

preserve the integrity of the American family, to foster and protect the viability of American family life by emphasizing family responsibilities in education, tax assistance, religion, and other areas related to the family, and to promote the virtues of the family (House of Representatives 614: 1)

the bill covered so many issues that it would have to have passed the scrutiny of at least five Senate committees. The 'fixers' in the Administration apparently did nothing to simplify what was in effect a 'wish list' rather than a viable piece of legislation. They also refrained from forcing the merger of two competing bills to permit public prayer in school. As might have been expected by experienced legislators, both bills failed to reach the floor of the Senate.

Even if a composite bill had not been passed, a roll call vote would have been useful to the NCR because a vote against the bill could have been used against liberal incumbents in subsequent elections.

For all Reagan's rhetorical support, his Administration appears to have been reluctant to commit its prestige and authority on the side of NCR legislation. In a letter to the editor of the *Moral Majority Report*, one NCR supporter commented on the failure of conservatives to mobilize enough votes to override a Senate filibuster of an amendment, introduced by Jesse Helms, to restrict significantly the availability of abortion:

> Where was the administration during this critical month-long battle to save the countless lives lost each year to the abomination euphemistically known as abortion? Inexplicably President Reagan, though calling abortion 'one of the most important issues of our time', was strangely silent on this measure during the month-long filibuster, and waited until the day before the first cloture vote on the filibuster to openly declare his support and fight actively for the Helms amendment. (Hohl 1982: 12).

If such sentiments were becoming at all widespread, then there was a serious commitment problem for the NCR, and it is perhaps the first evidences of that problem which were visible in the falling income to NCR organizations after 1984.[2]

The 1986 November elections gave control of both Houses back to the Democrats. The scandal over the illegal supply of arms to Iran and the syphoning of the profits to the Nicaraguan Contras considerably damaged Reagan's reputation. In spring 1987 Reagan chose to demonstrate that he was not a lame duck by vetoing a bill for massive expenditure on road improvements. Unfortunately for him, the bill contained so many local expenditure projects (or, to its critics, was so 'stuffed with pork') that even staunch Reaganites voted with their local interests to override the veto and pass the bill. In the autumn of 1987, the Senate voted 58 to 42 to reject Judge Bork, Reagan's nomination to the Supreme Court (of which more

[2] NCR activists put a brave face on their apparent decline in income and influence by arguing that (a) income always falls in 'off' election years, and (b) the GOP made a bad policy decision to neglect the NCR in 1986. Unpacked to the level of local campaigns, this means that they see the NCR's lack of impact as a result of conservative Republicans believing that they could win without raising socio-moral issues. As we cannot sensibly estimate the income of a movement as amorphous as the NCR, we cannot quantify the decline but the scale is indicated by the fact that in 1986 La Haye's ACTV spent only 10% of what it spent in 1984 (*Newsweek* 14 Apr. 1986).

later). With such defeats on more mainstream conservative interests, it was not surprising that there were no victories on the NCR agenda.

Compartmentalization, accommodation, and conflicting expectations are especially problematic because of the fragile nature of fundamentalists' commitment to politics. In contrast to the Protestants of Ulster or South Africa, who have a long history of involvement in political and ethnic conflict, and whose religious beliefs are a vital part of the dominant political ideologies, American fundamentalists have alternated between periods of pietistic retreat from the world and active involvement. In 1965, Falwell himself preached against pastors being active in politics:

> Believing in the Bible as I do, I would find it impossible to stop preaching the pure saving gospel of Jesus Christ and begin doing anything else—including fighting communism or participating in civil rights reforms. . . . I believe that if we spent enough effort trying to clean up our churches, rather than trying to clean up state and national government, we would do well. (in Bollier 1982: 71)

It is true that abstentionism was convenient for Falwell in 1965 because it gave him an apparently non-racial ground for criticizing Martin Luther King and other black pastors, but the convenience of the attitude made it no less genuine.

Even in Ulster, conservative Protestants are ambivalent about political involvement. Different elements of Protestant ideology lead to different conclusions about the necessity, value, and propriety of involvement. The individualistic element can be added to the belief that God has given us free will to sin if we wish to produce a willingness to let sinners get on with damning themselves. The Calvinist inheritance, far stronger in Ulster than in America, produces a willingness on the part of the saints to impose righteousness on the unregenerate. Yet one of the greatest benefits of the Reformation is political democracy and the essence of democracy is the freedom to choose the wrong as well as the right. Even the most highly politicized clergymen in Ulster—the ministers of Ian Paisley's Free Presbyterian Church (a number of whom are active in electoral politics)—are ambivalent about their right to impose their morals on others (Bruce 1986a: 125–34). It is hardly suprising that American fundamentalists with their roots in the Baptist 'free will' tradition should be less confident about political interventions, even without the long-standing American rhetoric of separating church and state. The oft-attended to and reported statements of the odd extremist who wants homosexuals imprisoned or women who have abortions hung have to be seen as precisely that: the views of the odd extremist. While keen to assert that they have as much right as anyone else to propose and promote their values, fundamentalists retain a considerable distrust for political activism. To cite just one of the many surveys which have demonstrated this point, only 45 per cent of a sample of Missourians who described themselves as 'born again' wished to see their pastors take a public stand on socio-moral issues. When asked about the more particular question of pastors endorsing candidates for elected office, only 25 per cent of the born again sample thought it acceptable, as compared to 16 per cent of the non-born again population (Patel *et al.* 1982: 258–60). As I argued earlier, such data have to be interpreted carefully. My discussions with fundamentalists suggest that, while they object to pastors openly stating a political preference, they fully endorse a pastor preaching on a certain passage of scripture in order to specify what sorts of public positions are Christian and hence to imply how the true Christian should vote. None the less, the general point is that fundamentalists are, as a consequence of their religion, prevented from being natural political animals.

The novelty and precariousness of most fundamentalists' involvement in politics always suggested (for those interested in looking for the signs) that failure to 'turn America round' would send them back to their verandas, back into pietistic retreat, and would make the Bob Jones University position of preaching revival and leaving God to do the social and political work increasingly popular.

I have argued that the NCR in America faces motivational problems essentially of two sorts: those which explain why groups other than Protestant fundamentalists have been less than enthusiastic in their support, and those which explain why fundamentalists, the core support, are not well suited to the role of political actor. The observations have been presented in such a way as to make it clear that the problems discussed are a natural result of trying to do religious politics in a pluralistic and secular society. This is the paradox which lies at the heart of the NCR and which is largely absent from the Protestant politics of Ulster and South

Africa. The encroachments of cosmopolitan secularizing society have forced American conservative Protestants to fight back. But fighting back, with any realistic chance of success, requires the adoption of some of the very attitudes which fundamentalists find objectionable.

However, before the above argument is taken as the obituary of the NCR, two observations should be made. The first is that the movement is not entirely dependent on continuing grass-roots support, although the disappearance of a considerable part of such support will have serious consequences. As resource mobilization theorists have argued, the structure of modern societies is such that a skilled cadre, provided it has the money, can present the appearance of representing a large body of public opinion and can continue to exert political influence. But, unlike Britain, America has elections for so many offices, and has major national elections so frequently, that popularity is often put to the test. Any suggestion of a reverse in momentum quickly affects the perceptions of those whose reactions produce a movement's impact.

In politics the relationship between apparent popularity and impact is more logorithmic than arithmetic. Those people, elected representatives for example, who depend on public support tend to overreact. Their responses amplify the importance of what it is they are responding to so that the process becomes a self-fulfilling prophecy. The initial impact of the NCR was massively exaggerated and produced an inappropriate fear of the right. Any sign that NCR support is weakening will be taken as a signal that socio-moral issues are off the agenda and the liberals can come out of the closet again.

This was precisely the reaction of many people to the results of the 1986 mid-term elections, in which all Ronald Reagan's best efforts could not prevent the Democrats increasing their majority in the House of Representatives and gaining control of the Senate. Socio-moral concerns were largely absent and, where conservative candidates did try to attack their opponents on new Christian right issues, they lost. Paula Hawkins, a Republican with a 100 per cent morality rating from Christian Voice, lost her Florida Senate seat. Linda Chavez lost in Maryland to an avowed liberal. A hero of the NCR, retired Admiral Jeremiah Denton of Alabama, was unseated. This does not, of course mean that the candidates in question lost because of their support for NCR issues. Just as British Conservatives could claim that they lost seats in Scotland in 1987 because Scotland had not had enough Thatcherism rather than having had too much, NCR activists could insist that their favourite sons and daughters lost because they were not sufficiently strident on the moral agenda. In the absence of good poll data, it is hard to estimate the 'real' popularity of NCR positions correctly. But it is not the 'reality' which is important; it is the perception, and the general perception—that developed by political analysts and presented in the media—was that the NCR was no longer a serious electoral force.

The apparent loss of momentum in late 1986 and 1987 was compounded by a major blow to the public prestige of fundamentalism which, although it resulted from the character weaknesses of only one individual, gave new life to a stereotype which has long dogged the whole evangelical culture.

SCANDAL AND THE STANDING OF EVANGELICALISM

Although television evangelists have played a major part in the rise of the NCR, their contribution is not entirely without its own dangers. They are polarizing figures, held in high regard by their devoted followings and disparaged by the rest of the population. Ironically a quality which was useful to the NCR may also have proved seriously damaging. Fundamentalist pastors enjoy considerable autonomy. Even those televangelists who are ordained ministers of large denominations are hardly supervised or guided. Their independence gives them the freedom to become involved with movements such as the NCR but it also means there are few curbs on their idiosyncracies. There is always the danger that a patron may turn into a liability.

James Robison is a popular Texas fundamentalist preacher who has been a considerable presence in the new Christian right in the South West.[3] In contrast to the avuncular Falwell, Robison is a pulpit thumper who uses hyperbole and coarse humour for effect.

[3] The James Robison Evangelistic Association produces a weekly syndicated television show and a magazine *Life's Answer*. In 1979 and 1980 James Robison produced a number of documentary-style programmes on political and socio-moral issues and he was active in launching Ed McAteer's Religious Roundtable. He also organized the 1984 Dallas National Affairs Briefing which gave a useful platform to then-candidate Reagan. His wife leads Christian Women's National Concerns Conference, a major Dallas–Fort Worth NCR group.

The following aside is typical of his style: 'As far as I'm concerned a homosexual is in the same class with a rapist, a bank robber or a murderer. You don't have to trouble yourself about whether it is normal to be a homosexual. God didn't create Adam and Edward. That's just not the program' (in Martin 1981: 223). Robison's reward for such views was an invitation to pray at the 1984 Republican nominating convention in Dallas.

One of Robison's more interesting acts was an episode of genuine (rather than metaphorical) iconoclasm. Texas millionaire Cullen Davis was converted under Robison's ministry.[4] After becoming convinced that it was sinful to own a collection of oriental art worth at least a million dollars, Davis gave the treasures to Robison. Robison had intended to sell them and use the money for his ministry but reading a passage from Deuteronomy—'You shall burn the carved images of their Gods with fire. You shall not covet the silver or gold that is on them'—led him to take the idols back to Davis. Together they smashed them and dumped the pieces in a lake (*Fort Worth Star-Telegram* 10 January 1983).

When I first mentioned this story (Bruce 1983: 22) I offered it, not as something which had damaged the public reputation of the NCR, but as a minor illustration of televangelists' potential for publicity-attracting eccentricity. I also then mentioned the more extravagant claims made in fund-raising, including the 900-foot luminescent Christ seen by Oral Roberts. In 1987 Roberts again attracted derision with his fund-raising, this time announcing that unless he got $6 million by a certain date, God would 'call him home'. However, my tentative prophecy was more than amply fulfilled by the fall from grace of Jim and Tammy Faye Bakker.

Jim and Tammy Faye Bakker were the hosts of *PTL*, a religious television show whose initials stood for Praise the Lord and People that Love. Before setting up on their own, the Bakkers had worked for Pat Robertson's *700 Club*. From their base in Charlotte, North

[4] In 1976 Cullen Davis was charged with the murder of his stepdaughter and the lover of his estranged wife, who was seriously injured in the attack. He was acquitted by a hung jury. He was later charged, after he was found with a gun, silencer, and $25,000 in cash on his person, with hiring someone to murder the trial judge (*Fort Worth Star-Telegram* 10 Jan., 18 Feb., and 7 June 1983). Davis and his present wife Karen (co-director of Christian Women's National Concerns Conference) are celebrities on the evangelical lecture circuit. Of no significance in his own right, Davis does demonstrate the tension between the novelty value of notoriety, which may give televangelists needed publicity, and the NCR's need for an image of probity.

Carolina, they had built a major cable television network and created a Christian theme park—Heritage USA—which was second only to Disneyland in the number of visitors it attracted. Although not as active in the NCR as Falwell, Robison, or Robertson, the Bakkers had frequently endorsed conservative political positions on their show and had given platforms to such conservatives as Jesse Helms.

For many years, the *Charlotte Observer* had run stories critical of the Bakkers' extravagant life-style. During one of their many financial crises, at the same time as appealing for sacrificial giving from their television audience, the Bakkers were spending $375,000 on an ocean front condominium and $22,000 on floor-to-ceiling mirrors. Such tales of extravagance could be ignored by the faithful but sexual indiscretion could not. In March 1987 Jim Bakker resigned from the PTL nightwork and asked Falwell to move in as caretaker chairman.[5] The reason for his departure was a story the *Charlotte Observer* was about to print which alleged that, in 1980, Bakker had drugged and seduced Jessica Hahn, a church secretary from Long Island. He had then compounded his indiscretion by arranging for the network to pay Miss Hahn some $265,000 to remain silent.

Once one skeleton had left the closet, others quickly followed. The wife of popular country and western singer Gary Paxton claimed that Tammy Faye's infatuation with her husband had played a major part in the break-up of their marriage. A magazine published a photograph of Tammy Faye and Karen Paxton, each dressed in what the normally sober *Washington Post* called 'a sexy dance-hall corset with a garter on one leg' (7 April 1987).

The problems of one evangelist quickly led to the questioning of the probity of others as Bakker turned first on his rivals and then on his friends. Although he admitted the seduction of Hahn, he claimed that the matter had only come to light because Jimmy

[5] What makes Bakker's later-regretted invitation to Falwell to rescue the PTL network a little surprising is the considerable difference in their theologies. With the exception of Robert Schuler, who offers something very close to Norman Vincent Peale's power of positive thinking, all the major televangelists are conservative Protestants. But Pat Robertson, Jimmy Swaggart, Jim Bakker, and Oral Roberts are pentecostalists who believe in the present power of the charismata of speaking in tongues and healing. Falwell is a fundamentalist who believes that these gifts of the Spirit are no longer readily available and that God now speaks primarily through the Bible.

Swaggart, a television evangelist from Baton Rouge, Louisiana, was trying to take over the PTL network. And three months after Falwell had assumed the task of saving the organization from bankruptcy, Bakker used an appearance on one of America's most popular chat shows to turn on him, claiming that Falwell had taken advantage of him in order to gain control of the cable and satellite network. When Bakker talked of coming back to the ministry, Falwell responded by calling a press conference during which he claimed that Bakker 'either has a terrible memory, or is very dishonest, or he is emotionally ill' (*Time* 8 June 1987: 44). Falwell also produced sworn statements from men who claimed to have engaged in homosexual acts with Bakker. Finally, he produced for inspection a note written by Tammy Faye Bakker on her own stationery in which she had outlined for her negotiator what she wanted from *PTL* as an incentive to retire gracefully: $300,000 a year for Jim and $100,000 a year for herself; all royalties from *PTL* records and books; the $400,000 mansion; two cars; security staff; and the fees for the lawyers and accountants they would need to protect them from the Internal Revenue Service. Even Falwell, who is by no means poor, found it difficult to hide his distaste for the arrogance and greed displayed in such a demand.

Jim Bakker and his former senior executive were expelled from their ministries in the Assemblies of God. Although Bakker must wait two years before applying for reinstatement to the ministry, he declared his determination to regain control of the network, hired one of the most extravagant lawyers in America—Melvin Belli—and started a number of law suits against Falwell. In October 1987, having failed to restore the organization's income to the level required for solvency and facing increasing problems with falling donations to his own programme, Falwell resigned.

There is no doubt that conservative Protestantism will survive the Bakker scandal (although it is not clear that the *PTL* and its network will). Fundamentalists and pentecostalists give considerable loyalty to their leaders but what unites them is their shared commitment to a body of doctrines and practices. However, Bakkers' problems do have general implications for the new Christian right. Most immediately, they caused the media to turn on Pat Robertson precisely the sort of publicity he did not want. While he was trying to present himself as a conservative business man who just happened to be a television evangelist, the press were interested only in his television evangelism and in his identity as the man who gave Jim Bakker his start in television. More long term, the scandal raised again the spectre of Elmer Gantry. Since 1979, pastors such as Falwell and La Haye have been gradually winning a place in the centre of American political and cultural life. Liberals and mainstream churchmen do not like them, but by 1986 even critical organs were giving fundamentalist spokesmen serious and thoughtful coverage. Television evangelists had ceased to be a joke. Suddenly even sympathetic conservatives found it hard to resist criticizing the Bakkers and, perhaps more damaging, laughing at them. In pointing to the business rivalries between the televangelists, one commentator said: 'Peccadilloes may be regarded as no big deal, but when you're talkin' hostile takeover, you're talkin' Sin!' (Safire 1987). With all the major mass media running features on the story, public attention was once more focused on the finances of mass evangelism. Every reporter found at least one sick old age pensioner living on welfare who regularly sent in a $10 cheque so that the Bakkers could have gold taps in their bathroom. Hardly suprisingly, the major television evangelists reported a decline in donations of some 30 per cent.[6]

The Bakker scandal and the many suits and counter-suits which have now been filed have done considerable damage to the reputation of the men who had begun to lose the 'holy roller' tag. The consequence for the commitment of the faithful will probably be no more than a shift in loyalty from one figure to another. What is more significant is stark confirmation of the negative stereotypes held by much of the general public. *Time* awkwardly but insightfully subtitled one story 'A sex-and-money scandal tarnishes electronic evangelism' (6 April 1987: 38). Anything which undermines the public standing of evangelical religion also undermines the appeal of the new Christian right.

CONCLUSION

In the final chapter, the weaknesses of the NCR which have been discussed here will be located in the broader theoretical consideration of the place of religion in modern societies. There I will identify those features of modernity which prevent movements such

[6] Summaries of the details of the *PTL* scandal are given in *Time* (6 Apr.; 4, 11, and 18 May; and 8 June 1987) and *Newsweek* (6 Apr. and 6 July 1987).

as the NCR achieving their goals. This chapter, and the previous one with its rather negative assessment of the impact of the NCR, very obviously imply that the NCR, in its present form, has not been a success and I have suggested some of the reasons for that failure. However, at the start of the chapter I cautioned against too early an announcement of the movement's demise. It is not unknown for movement organizations to persist long after most outsiders have written them off. The reverse is also the case: despite its failure to achieve lasting success on any of the key items on its agenda, the NCR, by shifting the centre of American public discourse, may have set in train changes which will posthumously further its agenda. The next chapter will consider this possibility in an examination of the NCR's relationship with the Republican party and the composition of the Supreme Court.

7

Kings and Judges: The GOP and the Courts

THE earlier discussion of the impact of the new Christian right demonstrated the importance of political parties and the judiciary for the concerns and actions of the NCR. At the end of the last chapter I suggested that the failure of the NCR as an effective social movement would not necessarily mean that none of its goals would be attained. This chapter will consider the recent past and likely future of the NCR's relationships with the Republican party, and changes in the judiciary, to reflect on the extent to which the agenda of the NCR might continue to be advanced after the movement's demise.

THIRD PARTIES, THE REPUBLICAN PARTY, AND THE NCR

The new Christian right has a great deal in common with previous third party movements in American politics and there are simple but important lessons to be learnt from their fate. The nativist movements of the nineteenth century, such as the 'Know Nothings', the American Protective Association, and the Ku-Klux-Klan of the 1920s were not enduringly successful. Certain problems were idiosyncratic productions of the movements in question but others were a natural part of the situation of movements which are mobilized around a single issue. Initially, their appeal is restricted to those who share their particular enthusiasm. Although many activists are quick to see the advantage of broadening the base of their movements by forming alliances with other single-issue groups, such alliances create tensions. The failure of the various populist movements of the 1890s—the Greenbackers, Goldbugs, Silverites, and poor farmers—to sustain a working coalition is an example (Canovan 1981: 20–51). Not only are there conflicts of interest as groups compete to feature their own particular enthusiasm or analysis of the cause of the wrongs which they intend to right, but there is often competition for leadership positions. To take an example from the 1930s, three populist movements combined to oppose the re-election of F. D. Roosevelt: the Townsend pension plan movement, the anti-communist movement led by Father Coughlin, and the remnant of Huey Long's Lousiania-based Share-Our-Wealth movement which was led for a short time after Long's death by Gerald K. Smith (Lipset and Raab 1978: ch. 5). Although there were major ideological differences which worked to subvert the union, it was summarily undermined by the inability of Townsend, Smith, and Coughlin to decide who should lead. With each unwilling to give way to the others, the Union party, as it was called, backed a little known state governor, William Lemke, who made a poor showing against FDR.

Third parties and politically interested social movements suffer a severe disadvantage in competing with the two main parties which pre-exist them. The single-issue movement may try to broaden its base by uniting with other such movements or by extending the range of its concerns. However, if this goal extension follows any obvious pattern of class or status related issues—that is, if it produces a package which is general and ideologically consonant—it will create something similar to what is already being offered by one of the major parties. To maintain a clear *raison d'être*, third-party movements must promote some cause which the main parties are neglecting. The need to maintain a distinct 'product profile' *vis-à-vis* the two majors forces any third party to amplify its unique features. Moderation will undermine the movement.

Social movement theorists have often neglected competition between movement organizations. With their interest in elections, the importance of competition has been more clearly recognized by political scientists. Third-party movements thrive in periods when the major parties are weak. To the extent that they are successful, their success is also their defeat. The pattern which one see' repeated over and over is of the initial growth of the third part being brought to a sudden halt and being followed by a rapic collapse as its most popular measures are taken over and presentea in more moderate form by one (and sometimes both) of the major parties. Whenever third-party movements were unable to gain enough electoral momentum to have significant impact on elections in their own right, they had to offer their support to candidates of the two main parties. Although some candidates were sufficiently honest and grateful to acknowledge the support of the Anti-Masons, the 'Know Nothings', or the American Protective Association, most, once elected, had very little to lose and much to gain by repudiating the third-party movements which had supported them. Lipset and Raab rightly point to the frequent phenomenon of 'defeat in victory' (1978: 144). Each wave of nativism ended with a major party accepting the more popular policies of the third parties, which failed: 'because the Whigs and their Republican successors were so well equipped for coalition politics that they could fit sizable amounts of conflicting single-mindedness and bigotry in their commodious national packages. . . . Their goals were partly taken over by a regular party . . .' (1978: 59). The taking over of goals often involved considerable watering down but it was enough to satisfy the more moderate supporters of nativist movements, who defected leaving them dominated by the extremists whose extremisim further alienated the public and hastened the movements' decline.

On other occasions, the third-party goals were not so much watered down as given little more than lip-service. In the run-up to the 1896 elections, the Democratic party very neatly undermined the Populists by nominating William Jennings Bryan, a well-known Silverite, and Arthur Sewall, for President and Vice-president respectively. While Bryan had good populist credentials, Sewall was an eastern banker and railroad director, 'a plutocrat whose only redeeming feature was his support for silver' (Canovan 1981: 45). In disarray, the Populists decided to support Bryan but to run Tom Watson for Vice-president. Although it lost the election, the Democratic party survived. The Populists did not.

Something similar has occurred with the NCR. If there is a difference it lies in the apparent willingness of the NCR to collaborate in the take-over process. Every early pronouncement stressed the non-political and bipartisan nature of the movement. In the event little effort was made to cultivate the Democrats. In 1924, a Klan delegation attended the conventions of Republican and Democratic parties and attempted to influence both platforms (Rice 1972: 70–80). Although Phillips, Weyrich, and Viguerie were more closely associated with Republicans than with Democrats, they all stressed the bipartisan nature of the movement they had helped mobilize. However, with the exception of one or two southern Democrats (such as Georgian Larry MacDonald) the Democratic party was not interested, and the NCR became firmly associated with the electoral fortunes of Ronald Reagan and conservative Republicans. And, as with previous third-party movements, there has been a distinct lack of gratitude from those who benefited from NCR support. A number of conservative victors played down the part the NCR played in their success. Although a number of NCR activists have been given positions in the Reagan Administration, they have generally been lowly offices.[1] Others have been absorbed by the Republican party organization. But despite the slim rewards, NCR leaders have been willing, almost keen, to be co-opted.

In part, this 'entrism' can be explained as the natural result of ambition on the part of the activists involved. A position in the GOP (the abbreviation for the 'Grand Old Party'—the nickname of the Republican party) is more prestigious and apparently closer to the decision-making centre than is working for an NCR lobbying organization. However, movement into the GOP also confers legitimation on the *values* which such activists hold. Closer ties to the party increase the chance of placing NCR issues on the legislative agenda. But perhaps more important than either of these considerations was simply the fragility of the early 'non-political' rhetoric. As Falwell recognized in his explanation of the mutation of Moral Majority into the Liberty Federation, the movement had found itself involved in an ever-wider range of issues, many of which could not be legitimated by the initial statement of purpose.

In the course of the past seven years, we have found ourselves drawn into issues and conflicts which were not anticipated in 1979. We have defended the Stragetic Defense Initiative. We have opposed an immediate unverifiable nuclear freeze. We have supported the Balanced Budget Amendment. We have supported financial aid for the Freedom Fighters in Nicaragua. We have openly opposed possible Communist takeovers in Taiwan, South Korea, the Phillipines, South Africa and all over the world. We have helped sponsor the Jerusalem Bill in order to move our embassy to Jerusalem. Many persons have felt that the Moral Majority name and charter are not broad enough to cover many of these domestic and international issues. We have therefore spent months discussing this problem with our national and state leaders. The end result has been the formation of a new organization which we have named The Liberty Federation. (1986: 1–2)

[1] It is noticeable that the very few members of the Administration—Lyn Nofzinger and Gary Bauer, for example—who are claimed as NCR men were committed Reagan supporters before they were involved with, or endorsed by, the NCR.

Although the additional issues had some bipartisan support, they were much more identifiably congenial to Republicans than to Democrats.

A further illustration of the degree to which the NCR has become a wing of the Republican party is found in the movement's relationship to election candidates, especially those in presidential races. In 1980 many NCR leaders avoided publicly endorsing Reagan (although they made it clear they found him more congenial than Carter or Anderson). By 1984 such reticence was very rare and many NCR leaders attended the Republican nominating convention in Dallas. Endorsements in 1980 and 1984 tended to be a matter of stating reactions to candidates already selected. Later endorsements took place before selection and were intended to play a part in the selection process; in 1986 Falwell publicly endorsed George Bush's ambitions to succeed Reagan. Subtle promotion and then overt endorsement of the GOP candidate have now been followed with an NCR candidacy; Pat Robertson is running for the Republican nomination in 1988. To the consternation of other men of the right such as Jack Kemp, Robertson has been polling surprisingly well in informal tests of Republican opinion. He will gain enough support in the primaries to force the GOP to feature a number of NCR issues in its platform and he may even have enough delegates to trade with the front runner for promises of future influence. If the party wins, the NCR may be well placed to demand some return on its support.

Although the federal campaigns attract the most attention, the NCR is also keen on having Moral Majoritarians selected for local office elections. In launching the Liberty Federation, Falwell added: 'we will also be challenging many of our people to run for office at the local, state and national levels' (1986: 2). The more realistic strategists believe that a solid base can only be built slowly with their novices acquiring experience at the lowest levels:

When some pastor decides he is going to be a Congressman that doesn't usually succeed, but a lot of these people are running at the local level. A charismatic and this other lady who is a Bible Baptist got together and decided to take over the Seattle school board. There is two ladies with zilch political experience and barely two nickels to rub together, they went out and took over the school board system of no small city. . . . We have to think long term . . . The local level is where you can get trained as a political leader. Maybe a few terms on the school board, then you can run for city council and maybe after a couple of terms on city council, you can run for county commissioner or state legislature or something like that.

Coupled with this strategy is gradual advancement in the GOP structure. The hoped-for consequence is that conservative Christians become an established interest group within the party, in much the same way as the racial minorities represented by Jesse Jackson have sustained their claims within the Democratic party. The parallel is a useful one. Although it has done far better than many commentators expected, the Jackson campaign has been hindered by the heterogeneity of the elements in his 'rainbow coalition'. Similarly, the various fissions and tensions within the NCR make successful caucusing difficult and suggest that the considerable presence of socio-moral conservatives in the country as a whole will not easily turn into a strong bloc within the party.

Just as happened with nativism, the co-option of the NCR by the Republican party has, for the NCR, the positive consequence of giving considerable legitimacy to its positions. But acquiring the status of an established faction within the Republican party (if that can be done) hardly increases the NCR's chance of significant success when the House of Representatives continues to be controlled by the Democrats—as it has been since 1954—and the Senate is now back in Democratic hands, after the brief period of Republican control from 1980 to 1984. Considering 'mega-trends' in political analysis, what is important is the likelihood of major realignment. If conservative southern Democrats—the Dixiecrats—join the Republican party, one would have in America what one has in most European countries: a party of the right and a party of the left.

Some southern Democrats have changed parties. Strom Thurmond, who stood for President on the Dixiecrat anti-civil rights ticket in 1948, now sits in the Senate as a Republican. Jesse Helms was a registered Democrat before he became a candidate. The civil rights movement created a major reaction among southern Democrats who temporarily abandoned the party to support the independent campaign of George Wallace, and some of them moved to the Republican party. Most did not. Wallace himself courted working-class black voters so well that he came to poll better in black districts than he did with white voters. Most

Democrats stayed with the party and the long-awaited major realignment did not happen.

As the more astute political commentators have noticed, there is no necessary connection between conservative positions on socio-moral issues and race relations, and economic conservatism. It is not just the historical accident of party positions on the Civil War which explains why the old South votes Democrat. For all that views on civil rights and race are a major source of divison between southern and northern sections of the Democratic party, the two parts share a common liking for 'big government'. As Phillips notes in his analysis of attitudes to government economic intervention: 'a conservative government resting on a Sun Belt base stands a better chance of moving towards the views of John Connally than of Adam Smith—toward pork barreling, crop subsidies, agribusiness, high technology and aerospace mercantilism rather than fidelity to free-market principles' (1982: 97).

The whole question of realignment is too complex to be considered in detail here (but see Lawrence and Fleister 1987). It is sufficient to appreciate that the major realignment which was predicted for, and sought by, Richard Nixon was halted by the fall-out of the Watergate scandal. Again in 1980 and in 1984, a major consolidation of American conservatives under the leadership of Ronald Reagan was predicted and did not appear. Most thoughtful commentators have stopped trying to interpret every election in terms of simple realignment and have instead come to the conclusion that one is seeing 'dealignment'. There no longer appears to be the party loyalty that once characterised electoral choice. More and more voters are 'splitting the ticket'. Although the simplest response to the very long ballot papers which face many American voters in elections is to vote the party slate, it is becoming increasingly common for voters to vote for one party for presidential and congressional office and for the other for state and local office. To give the example of the California electorate in November 1986, voters re-elected the conservative Republican Governor but rejected his ideological soul mate for lieutenant governor. They threw out liberal Justice Rose Bird but re-elected liberal Senator Alan Cranston. In South Carolina, an incumbent Democrat was re-elected to the Senate, a Republican won the gubernatorial contest, and a Democrat was elected lieutenant governor (Ornstein 1986: 15).

Although it would be foolish to predict the future ideological thrust of the Republican party, it is already clear, in the aftermath of the 1986 mid-term elections, and in the reactions to the 'arms for hostages in Iran' scandal in the Reagan Administration, that more mainstream Republicans such as George Bush, Howard Baker, and Bob Dole, rather than 'movement conservatives', are the likely standard bearers for the GOP in the foreseeable future. And, with no sign that the Republicans are becoming the new majority party, the party's leadership hardly matters.

The common occurrence that Lipset calls 'defeat in victory' is of little advantage to a social movement if the major party that is stealing the clothes fails to come to power. The absorption of the NCR into the Republican party may be of some long-term help in that it should alter public perceptions of the NCR's goals. Already Reagan's public support for school prayer, anti-abortion campaigns, and creation science has helped to shape reactions: what were once the ravings of religious fanatics are now the serious issues of some right-thinking conservatives. The Republican party's endorsement of such positions will help to establish them as part of the legitimate ideological currency of the right. However, there is a considerable difference between making certain values sufficiently respectable so that their carriers are no longer laughed at, and actually passing laws which promote them. While the returns from acting as a faction within the GOP will probably not be so poor as to justify renaming the process 'defeat in defeat', they will be small.

A NEW PARTY?

Although it is at present only a gleam in the eyes of one or two strategists, an interesting variant on the third-party idea is attracting the interest of some in the new Christian right. Without exception, every NCR activist I interviewed in 1986 and 1987 admitted that the movement had been poorly rewarded by the Republican party. Although most saw continued effort to take over local parts of the GOP structure as the best strategy for the future, some floated the idea of a third party which would avoid the mistakes of earlier attempts. Those mistakes were quite correctly identified as (a) reliance on a single issue, (b) constructing a party around a single individual (as in the Wallace campaigns), and (c) fielding candidates when the new party was unlikely to avoid an embarrassingly bad result. One suggested way of avoiding these problems is to build on the experience of the New York Conservative party which acts as a 'surrogate'. It has its own membership, ideology, and structure but does not usually field candidates. Instead it acts as an additional primary, 'external' to the Republican party. It makes it clear what positions it will support and endorses (or refuses to endorse) the candidates of the Republican party. A new conservative party, built on the infrastructure of Moral Majority Inc./Liberty Federation, Christian Voice, Religious Roundtable, and other NCR organizations, would similarly vet GOP candidates. The third party would bind together the sorts of constituencies which presently support the NCR. It would give them an organizational structure in which they would feel more 'at home' and in which they would not be shunned or ridiculed. They could open and close their meetings with hymns and prayers if they chose, without feeling that they were there only on sufferance. If they were sufficiently powerful and could deliver enough campaign workers, money, and voters, in many areas the GOP would be tempted to tailor its candidate selection to the wishes of the third party.

While a surrogate third party might increase the NCR's influence on the GOP, it is questionable whether such a scheme will leave the drawing board. A sober consideration of survey data or indices of electoral impact does not suggest that the NCR has the ideological coherence or the commitment necessary to make a third party work. The NCR does not have the single-mindedness of a single-issue group (such as the anti-abortion movement). But to the extent which it moves in the other direction and becomes an extremely broad conservative movement, it comes to compete with the Republican party, a competition it must lose.

JUDGES: THE FEDERAL COURTS

As I made clear in Chapter 2, the federal courts have been a major vehicle for the promotion of the modernizing and secularizing influences which conservative Protestants find so objectionable. In explaining the sense of grievance which led many conservative Protestants to feel that political mobilization was justified and necessary, I pointed to the general centralization of decision-making and the gradual erosion of the autonomy of the regions. The courts played a major part in that centripetal process.

Most conservatives object to the judicial activism of the post Second World War Supreme Court.[2] Although the courts are limited in their actions by a document which lays great stress on the rights of the individual, they have none the less created a considerable body of liberal and progressive law. But apart from a general dislike of what the courts have done in the past, the new Christian right has an additional reason for being critical of the judiciary; the federal courts are an obstacle to future progress on the NCR agenda. The issues which concern it (such as gay rights or abortion) are either in some sense fundamental or involve the correct relationship between church and state. Hence, as the fate of the Arkansas and Louisiana 'equal time for creation science' bills demonstrates, local successes of the NCR come to be weighed in the balance by the court system: state and federal, and then, often, by the Supreme Court.

The NCR (and the right generally) has consistently argued that America faces a serious constitutional crisis in which the judges have become too powerful. The promoted alternative is 'judicial restraint': the view that the bench should confine itself to a narrow and traditionalist interpretation of its role and not engage in the sort of activism which characterized the Warren and Burger Courts. Conservatives accompany their plea for restraint with a demand for a return to 'original intent'. As Reagan's Attorney General Edwin Meese III has argued, it is not for the Court to make new law. That is the place of Congress. The Court should confine its constitutional role to the judgement of the fit between present judicial decisions and the 'original intention' of the framers of the Constitution.

> In particular, judicial restraint would involve a rejection of: the doctrine of incorporation in the Fourteenth amendment by which guarantees of the Bill of Rights like freedom of speech, press, association, religion, privacy, counsel and freedom from excessive punishment have been made binding on state and local officials. (Schwartz 1985: 5)

In the early stages of the evolution of the American polity, the Bill of Rights was taken to limit only the power of the federal

[2] There were considerable differences in the activism of the Burger and Warren Courts but these need not concern this study. Conservative Protestants (and conservatives generally) rarely make such distinctions; for them, all post-war Courts and Court decisions have been bad.

government. In the modern period, most justices have followed the centralization route. These 'incorporationists' argue that, because of the Fourteenth Amendment, the same principles which govern the relations between the federal government and the individual also govern the relationship between each state and the individual. What the federal government cannot do, the states cannot do.

In the battles over civil rights, southern segregationists returned to the old states' rights platform and objections to incorporation have again been raised by conservative jurists in their writings on the socio-moral issues which concern the new Christian right. Those who object to the Court's construction of church–state relations in the First Amendment have tried two strategies. One is to amend the Constitution. An amendment permitting public prayer in schools reached the floor of the House of Representatives in 1971, but failed narrowly to get the two-thirds votes needed to send it to the states for ratification.

Efforts in the 1980s, led by Senator Helms of North Carolina, have tried the different route of attacking the doctrine of incorporation. Helms introduced a statute which would have limited the jurisdiction of the United States Supreme Court and the lower federal courts so that they would not hear cases involving 'voluntary prayers in the public schools and public buildings'. What Helms is suggesting is the removal of the system which imposes the cosmopolitan values of the centre of American government and the freeing of the states to pursue their own preferences. Although such liberty would lead to the infringement of the rights of minorities in states where there was widespread support for public prayer, it would permit the reintroduction of prayer simply because the degree of pluralism in any state would be lower than it is for the country as a whole. The Helms plan has to date failed to win congressional support. Although the only significant legislation along these lines which has been proposed concerned school prayer, the handing back of issues of religious particularism to the states would benefit the NCR agenda generally. The return of local autonomy would make it impossible for liberals to use the federal courts to prevent state legislatures passing 'equal time for creationism' bills. It was on the ground that such a result would be closer to the 'original intent' of the drafters of the constitution that Justices Rehnquist and Scalia dissented from the majority verdict on the Louisiana equal time bill.

Although the NCR objects to the relative unaccountability of the federal courts, it is quite willing to benefit from it and, under the presidency of Ronald Reagan, it is likely to do so. Reagan has elevated William Rehnquist, the most conservative member of the Court, to the position of Chief Justice and appointed another conservative—Antonin Scalia—to fill his seat. The resignation of eighty year-old Justice Powell in June 1987 created a second vacancy which Reagan sought to fill, first with Robert Bork and then with Douglas Ginsberg, both confirmed conservatives and advocates of judicial restraint. Reagan has created a task force to suggest and vet federal court appointments in order to ensure that his appointees are not only able and extremely conservative, but also young. Given that judges do not have to retire, and that by 1988 almost half the federal judges will have been appointed by Reagan, the NCR has considerable hopes that future judicial decisions will favour their positions.

All has not been plain sailing, however. Although the nominations of Rehnquist and Scalia were not opposed by the Senate (which has to confirm senior judicial appointments), other Reagan appointments have been challenged. People for the American Way, the major anti-NCR lobbying organization, devoted considerable resources to stimulating opposition to the nomination of Daniel A. Manion for the 7th US Circuit Court of Appeals. Nominees must be confirmed by the Senate. They are cross-examined, often at length, by the Senate Judiciary Committee, which reports to the whole House. Although the Senate as a whole can override the recommendation of the Committee, disputes rarely come to a floor vote. Potential nominees are normally discussed informally first and, until the Reagan Administration, strong opposition, either from influential members of the Judiciary Committee or from organizations such as the American Bar Association, was usually enough to prevent a name going forward. However, once a nomination is made it is not usually opposed solely on ideological grounds. After all, the business is politics and a cardinal principle of politics in a system where parties alternate is that the members of the dominant party hesitate to do to others what they would not like to have done to them when they are in a minority. Even though one ideological group (these issues are usually ones for factions rather than parties) may find a particular nomination distasteful, it will bear in mind that it may soon have a 'favourite son' it wishes to reward with a seat on the federal bench. Thus, although the nominations of Rehnquist and Scalia were subjected to vigorous scrutiny, they were approved. Although the two men were extremely conservative, they were well qualified. However, liberals were able to contest the nomination of Manion on the grounds that, in addition to being a supporter of the John Birch Society and the son of Dean Clarence Manion—the host of a popular 1950s right-wing radio show, the *Manion Forum* (Forster and Epstein 1964: ch. 7)—he had publicly flouted previous Supreme Court judgements. After the Supreme Court in 1981 had struck down a Kentucky law which required the display of the Ten Commandments in all public school classrooms, Manion had co-sponsored an almost identical bill in the Indiana state senate. He could also be described as judicially inexperienced. The Senate Judiciary Committee divided and the nomination went to a floor vote where it was approved by a very narrow margin after heated debate.

Not deterred by the result, PAW put even more effort into opposing the nomination of Judge Robert Bork to the Supreme Court and this time the liberals succeeded. Not even his enemies denied that Bork was extremely able and he was not an outsider; he had served as Solicitor General under Richard Nixon. He was opposed simply because he was too conservative. In particular PAW managed to raise considerable public fears that Bork would vote to reverse many of the legal gains made by women and racial minorities. Considerable black opposition meant that many southern Democrats who by personal inclination might have liked Bork had to vote against him because they feared losing the black vote in their constituencies. The Judiciary Committee divided 9 to 5 against, with all the Democrats and some of the Republican members opposing the nomination. Instead of withdrawing Bork, Reagan took the advice of movement conservatives and forced the vote on the floor of the Senate. The vote was already lost but forcing senators to declare their hand would give conservatives something to beat the waverers with at the next election. Bork was rejected by 58 to 42 votes.

The response of most previous administrations would have been to accept defeat and nominate a safe, middle-of-the-road judge whose confirmation would be a formality. This is the course which Howard Baker, Reagan's Chief of Staff, suggested but Reagan chose instead to take the advice of Attorney General Meese and nominate another conservative who was quickly dubbed 'son of Bork'. In the event Ginsberg's nomination self-destructed when he admitted to having smoked marijuana. For his third nomination, Reagan accepted the advice Howard Baker had given the first time and nominated Anthony Kennedy, a popular non-doctrinaire conservative with a long record of reasonable judgements. The Senate approved the nomination unanimously and Reagan's chance to move the Supreme Court sharply to the right had gone.

One has to be a little cautious about predicting the consequences of court packing. There are many examples of supposedly conservative justices becoming centrist and even radical once appointed. Earl Warren, who had a moderately conservative record as Governor of California, was chosen by Eisenhower to reverse the liberal tide of the previous Court. Instead he presided over a series of reformist decisions on social and racial issues. Nixon appointed Warren Burger to reverse the liberal activism of the Warren Court. Instead, his Court upheld busing as a legitimate tool for desegregating schools. Reagan's first appointment—that of Sandra Day O'Connor, the first woman justice—has also failed to live up to conservative expectations. Although Rehnquist has been the lone dissenting voice on a sufficiently large number of judgements for there to be little doubt about his radical conservatism, the tendency of previous appointees to deviate from the political postures for which they were appointed, towards a position more in line with public opinion, should make us cautious of supposing that the packing of the federal courts with conservative justices will necessarily cause a radical shift to the right. It is because they are alert to this that officials in the Reagan Administration have shown a preference for academics who have stated and restated their conservative views over many years. Even though it is theoretical, such people have a clear record and are less likely to change their views.

But even with that caution the appointments already made, especially the two changes in the Supreme Court, should result in a slightly less aggressively secular interpretation of church–state relationships and a shift to the right in many other controversial areas. One can only speculate on just how much movement there will be on particular issues but it is very clear that the issues dearest to Protestant fundamentalists are least likely to be resolved in their favour. A recent Court decision not to permit a range of restrictions on the availability of abortion was decided 5 to 4 in the liberal direction. Clearly the appointment of a doctrinaire right-winger in place of Powell, who was often the 'swing vote', would have called into question the future of further legislation on abortion but the liberals were strong enough to block Bork. But the vote against the creation science bill was 7 to 2. Adding even two more conservatives will not change that. Finally, the recent judgements on school texts in Alabama and Tennessee (discussed in the last part of Chapter 5) suggest that the fundamentalist strategy of appealing to the Constitution and fairness to have 'non–fundamentalist' books excluded from the curriculum has run out of steam.

To what extent such changes in the ideological composition of the judiciary as have occurred can be attributed to the NCR is a difficult question. It may well be that Reagan was simply following his own conservative instincts and would have acted as he did without the NCR. The nomination of Bork, who was first brought into the political arena when Richard Nixon made him Solicitor General in 1973, would support such an argument. His reputation as a conservative jurist and his previous service in Republican administrations would have made him an ideal right-wing candidate for the Court even if the Moral Majority had never existed. If one leaves aside the NCR for a moment and simply considers the appointment of conservative justices, one sees merely continuity between Nixon's appointment of Rehnquist to the Court and Bork to the office of Solicitor General, and Reagan's nominations of Rehnquist for Chief Justice and Bork for the Court. Perhaps the most one can say is that the NCR, in popularizing and making more respectable conservative socio-moral positions, has reinforced the convictions of Reagan and his conservative advisors and made it easier for them to do what they probably would have done anyway. That Reagan could not get Bork onto the Court simply demonstrates the limits of NCR influence.

CONCLUSION

The absorption of elements of the new right and new Christian right by the Republican party and the likely slight shift to the right in federal court decisions on contentious socio-moral issues are two changes which have complex and as yet unclear connections with the success of the NCR as an organized social movement. They are treated together in this chapter because they share the feature that, while they can be read as minor successes with the potential for becoming greater victories, they also contain elements of defeat.

Being co-opted by the GOP may be seen as a victory in that it gives status to NCR activists and hence legitimacy to their positions. Falwell and other NCR leaders are extremely fond of reproducing photographs of themselves with President Reagan and other Administration officials. Almost every issue of *Moral Majority Report* carries lengthy reports of meetings between Moral Majority leaders and Administration officials and cabinet members. Such meetings are arranged and reported for the purpose of establishing the credibility of lobbyists. They serve a dual function of showing fundamentalists that politicians take their leaders seriously, and of showing the general public that conservative Protestant lobbyists are serious 'players'. However, politics invites cynicism, and a certain amount of it seems appropriate here. As Lyons in his comments on the willingness of the Arkansas legislature to approve bills which announce that Christianity is a good thing, and Neely in his more detailed analysis of legislators' electioneering, both show, rhetorical support for particular interests is easy and it is cheap. It cost very little for the Reagan Administration to invite fundamentalist pastors to breakfast and to declare a 'year of the Bible'. It cost very little for the Republican party to appoint NCR activists to very junior positions. It cost very little for the GOP to insert statements of support for conservative socio-moral positions in its presidential election year platform. It cost such mainstream Republican candidates as Jack Kemp little to thank Pat Robertson's candidacy 'for bringing in evangelical Christians' (*Time* 9 November 1987). While these represent important symbolic victories which the NCR can use to claim credentials as a serious political force, they are not actual legislative victories. And, given that there has not been the major realignment necessary to transform the GOP into the natural majority party, the symbolic gains for the NCR of becoming a faction within one of the established parties are offset to the extent that the party in question is the losing party.

Ironically, it is the change in the ideological composition of the federal court bench, that body of 'undemocratic and unelected' policy makers so frequently criticized by the new Christian right, which is more likely to serve the ends of the NCR. But even here we should be careful of accepting the rather hysterical liberal estimates of the consequences of such changes.

Firstly, it is worth appreciating that, right or wrong, the present Court position on church–state relations, which might be bluntly characterized as 'no nothing', is relatively easy to operate. Any move from this to allow the reintroduction of religion into the public sphere would destroy a position that works without providing a usable alternative. Some extremely conservative justices might be willing to re-examine the issue but many even quite conservative jurists see little point in muddying the waters.

The professional interests of the justices offer one good reason for doubting a lurch in the direction of the NCR. Another good reason concerns the Court's respect for public opinion. Neely is probably right when he argues that, for all that it is undemocratic in its creation, the Supreme Court and the lower federal courts generally reflect public opinion (1981: 10). As a rough 'rule of thumb' for distinguishing good from bad judicial activism, Neely suggests the fit between the decisions the courts make and the way Congress would have voted if the same issue had come to the floor and legislators had temporarily laid aside their reluctance to make decisions. On that test, all the crucial decisions made by the Supreme Court over the last two decades, with the exception of support for enforced busing of school children, were 'good' decisions. In good judicial activism, judges make decisions which the public want and which legislators know to be good decisions. The reason they have not made those decisions is that the vagaries of the structure of American politics does not always allow serving the public good to appear in the considerations of elected representatives as a sensible career move. Then there is bad judicial activism where judges make decisions which reflect their own attitudes but which are generally unpopular. These cases tend to be reversed. The conclusion of the supporters of judicial activism is that, although they are not directly responsive to external interests, in the long run the judgements of the courts tend to reflect public interests and concerns. If they are right, then the 'activism' of rightist judges may well be similarly tempered. With the NCR apparently running out of electoral steam, there will not be the rightward pressure on the courts that many critics of the NCR feared.

8

Revelations: The Future of the New Christian Right

THIS concluding chapter will consider the significance of the new Christian right, summarize its strengths and weaknesses, and speculate on the future of the sort of conservative Protestant politics it represents. One way of framing a picture of the NCR is to consider two words that often appear in explanations of it: revival and reaction (Peele 1984).

Revival or reawakening can be taken in two senses: as referring either to the Christian Church as a whole and thus to an addition of numbers to the body of the saved; or as referring only to the 'faithful', thus implying a new sense of vitality. Evangelicals, and people writing about evangelicals, usually use the term to suggest a period of widespread conversion, a time in which new members are being added to the churches. When used in discussions of the new Christian right, it suggests that the explanation for the movement lies in the increasing popularity of evangelical Protestantism. Many discussions of the NCR begin by placing the movement within the context of the decline of the mainstream Protestant churches and an increase (relative or absolute depending on the author) in the numbers belonging to evangelical, pentecostal, and fundamentalist churches.

Although there is something in the observation that the rise of the NCR follows the decline in the size and confidence of the mainstream churches, and hence is related to the increase in confidence of conservative Protestants, it would be misleading to imply that the politicization of some conservative Protestants was a *natural* consequence of an increase in their numbers. In so far as any connection can be discerned it is almost the opposite; the expansion of the milieu has been accompanied by the reduction of its distinctiveness and hence of its grounds for a distinctive politics. As the claim to be 'born again' has become more popular and respectable, the amount of behavioural change, of asceticism, associated with that state has been reduced. As Quebedeux signalled in the title of his book, *The Worldly Evangelicals*, the gradual reduction of the amount of world-rejection involved in 'getting saved' which John Wesley observed in his upwardly mobile followers is being repeated among American evangelicals. Hunter's study (1987) of young evangelicals shows a small but crucial change in attitudes; a relaxation in their own standards has been accompanied by an increasing unwillingness to condemn out of hand those who differ from them. Some of the certainty has gone and in its place there is an element of recognition of socio-moral pluralism. Although young evangelicals still have a strong sense of what is right for them, they no longer seem so sure that what is right for them is also right for everyone else.

There is a slight problem in interpreting these data. It might be that the expansion of conservative Protestantism has produced some relaxation at the peripheries because people who have always been conservative Protestants are less willing to forgo the pleasures which their increasing prosperity is now making available to them.[1] Alternatively, it could be that the expansion has meant the incorporation of newcomers who wish some of the rewards of being born again without making what were previously mandatory sacrifices. Either way, the increasing self-confidence of conservative Protestantism has been accompanied by a relaxation of standards. This incipient pluralism and moderation offsets much of the advantage of self-confidence. Conservative Protestants may have had their morale boosted by having Jimmy Carter, a born again Baptist lay preacher, as President, but he acted like a liberal.

The point will be pursued shortly but here it is enough to say that the simple connection of conservative Protestant revival and increased politicization assumes too monolithic a view of conservative Protestantism. A good part of conservative Protestant growth is associated with decline in the conservative elements.

The term 'reaction' is often used for two distinct purposes. Sometimes it identifies a general shift to the right in American politics; sometimes it identifies the NCR as a reaction on the part of a particular section of the American people. Taking the first possibility, the political mobilization of some conservative Protestants is seen as simply one element of a general shift to the right in

[1] This is the Niebuhr thesis (1962) for the common development of sects into denominations.

American politics and culture. In this scenario, Nixon's rout of McGovern and Reagan's victories in 1980 and 1984 are evidence of a general move away from the liberalism of the Kennedy era and Johnson's Great Society, and the election of Jimmy Carter in 1976 is the exception explained by Nixon's Watergate disgrace. The idea that America has shifted to the right is one that has widespread currency, even among liberals. Especially since 1980, a significant number of younger Democratic politicians have seen it as their task to reconstruct the party to accommodate this new conservative mood.

There is actually little clear evidence for a general and significant shift to the right. In a lengthy analysis, Ferguson and Rogers (1986) persuasively argue that opinion poll data from the 1960s to the present show little evidence for a rightward shift in public thinking. For example, on attitudes towards business, support for government regulation of company profits actually *increased* between 1969 and 1979. The percentage of people thinking that business as a whole was making 'too much profit' rose from 38 per cent to 51 per cent.

> Over the period 1971 to 1979 the percentage thinking that 'government should put a limit on the profits companies can make' nearly doubled, rising from 33 to 60 percent. . . . As the rollback in regulation and cutbacks in domestic spending became evident during Reagan's first term, the public increased its support for regulatory and social programs. (Ferguson and Rogers 1986: 44–5)

Nor is there any great evidence of a move to the right on socio-moral issues. An ABC poll found that support for abortion on demand, no matter what the reason, had increased from 40 per cent in 1981 to 51 per cent in 1985, while the percentage opposed to abortion on demand had gone down from 59 per cent to 46 per cent. On the issue of affirmative action, the Reagan era has seen a rise in the percentage of people declaring themselves in favour of a 'federal law requiring affirmative action programs for women and minorities in employment and education, provided there are no rigid quotas' (Ferguson and Rogers 1986: 46). Given the excellent opportunities for divergent readings of these sorts of questions, one would be unwise to use the considerable amount of poll data which Ferguson and Rogers have amassed to argue that Americans have become conspicuously more liberal, but it is certainly difficult to find any reason—except for the election of Reagan, that is—to suppose that they have become more conservative.

Two minor points should be made about Reagan's popularity. Firstly, even in his 'landslide' of 1984, he won only 59.6 per cent of the votes cast; 40.4 per cent went to Mondale. Gallup regularly records presidential popularity ratings. Over the four years of his first term, Reagan's rating was 50 per cent, which, as Ferguson and Rogers note, was *lower* than that of Eisenhower (69 per cent), Kennedy (71 per cent), Johnson (52 per cent) and Nixon (56 per cent) and not much above that of Carter (46 per cent). Reagan's peak rating was shortly after taking office when his performance was approved of by 68 per cent of those polled. Eisenhower (79 per cent), Kennedy (83 per cent), Johnson (80 per cent), and Carter (75 per cent) had higher peaks, while Nixon's 67 per cent was not significantly lower. The second point is that, as Ferguson and Rogers plausibly argue using correlations with economic indicators and poll data, Reagan's popularity and election victories are mostly simply explained by the performance of the economy. The weakness of the economy cost Carter the 1980 election and its strength in 1984 won Reagan a second term.[2]

Put briefly, there is little reason to suppose that there has been a major shift to the right. Furthermore, even those analysts who see a move to the right on economic issues add that there has been no corresponding shift on socio-moral issues. Haynes Johnson, a Pulitzer prize-winning journalist who has been writing about American politics for over thirty years, said:

> the real revolution in America is not a political but a personal one. By that, I mean a revolution in personal attitudes and values—sexual and racial primary among them. These changes in behavior and attitudes, startling when compared to the more straightlaced traditional America of a generation ago, also hold significant political implications. We have today, if I'm right, a citizenry that is far more liberal in terms of its social values—favoring abortion, much more tolerant about sexual behavior, about divorce, about racial relations and civil rights—and yet also notably more conservative when it comes to fiscal and governmental matters, especially government from Washington. (in Duke 1986: 39)

[2] Although I have questioned the construction of complex variables in their research, there is less reason to challenge the responses to straightforward questions in Johnson and Tamney's Middletown study. They claim that the item most frequently chosen as 'most important' from a list of issues was 'inflation', and that those who chose it voted 2 to 1 for Reagan (1982: 128).

Even those who are Reaganite on the handling of the economy, defence, and foreign policy wish to maintain the freedom to choose their own life-styles.

The second reading of 'reaction', which sees the NCR as a reactionary political movement supported by a section of the population which feels itself threatened, is more plausible. However, as I argued at length in the first chapter's discussion of status defence or status anxiety theories of collective behaviour, the notion of reaction has to be used carefully if it is not to mislead. We should not see the NCR as an almost neurotic response to social, political, and cultural changes which threaten small town Americans and their traditional values. Those analysts who have tried to deploy status defence models of the NCR have usually had to admit that they do not work (Simpson 1983, 1985; Buell 1983). Unless status is defined so broadly as to be meaningless, supporters of the NCR do not share a common status. Apart from a tendency to be located in the South or South West, the main thing that supporters of organizations such as the Moral Majority have in common is their religion. The NCR is a movement of cultural, rather than status, defence. To use the American phrase, it is concerned with the politics of life-styles rather than status.

The word 'reaction' also has the unfortunate suggestion of bad faith about it; it implies that, while liberals have authentic values, which they hold dear for good reasons, their opponents take up positions in reaction. The core of conservative Protestant values has remained remarkably stable for a long time. The social, political, and moral conservatism of evangelicals and fundamentalists is not new. They have not become conservatives in response to the increasing permissiveness of liberal America. They have stood still while all around them moved. If liberals see the conservative movement as a reaction, this is the illusion of children in a train which is leaving the station supposing that they are standing still while the stationary train next to them is moving backwards.

What is novel or reactive about the actions and attitudes of people like Falwell, Robertson, Dixon, and their followers is their willingness to campaign publicly and politically against things they find offensive. Although fundamentalists have always found something to complain about—after all this world is the world of fallen and sinful man—their environment has grown increasingly hostile. The unregenerate part of the world has become ever more obviously unregenerate. One need not follow fundamentalists in their uncritical attitude to the past, their blanket condemnation of the present, nor in their explanation of the ways in which the world has changed to accept that divorce is now common, as is drug addiction, that homosexuality is accepted in many circles as an alternative life-style, that 'housewife' is a devalued status, that the separation of church and state (once interpreted as denominational neutrality) is now taken to imply secularity, and so on. The changes which have been promoted and welcomed by atheists, feminists, racial minorities, and liberals are changes which have fundamentally altered the moral, social, and political culture of America and moved it away from the standards and practices which fundamentalists regard as biblical.

Furthermore, and probably more importantly, the changes in American culture have been accompanied by a social force which amplifies the offence to conservative Protestants: increased centralization. Although America remains far less centralized than Britain, the general trend has been in the direction of increased cultural and economic penetration of the peripheries and increased government intervention in the lives of individuals. I suggest that increased intrusion is more important than increased liberalism because it is the 'proximity' of what offends which defines the extent of the offence. Conservative Protestants always knew that people in Los Angeles or New York were desperate sinners but so long as they could live in their own shared and socially constructed subsection of America, it was not something that really hurt.

Some sub-societies and subcultures are geographical. They are regional peripheries, backwaters, or 'ghettos'. Others are social constructions which can, with effort, be sustained in the 'centres' of modernity. However, in either form, such subcultures depend on structural conditions permitting their survival. Conservative Protestants felt themselves pushed into political action because they saw the state making increasing demands on them. They used to have racially segregated schools which began the day with public prayers. Racial integration is now public policy and the courts enforce a strict separation of church and state. Conservative Protestants react by establishing their own independent Christian schools. The state then intervenes again, depriving those which appear to be segregationist of their tax-exempt status and supporting those state legislatures which require the licensing even of independent schools. It is this sense of having their autonomy reduced which explains the considerable hostility to cosmopolitan America. This point is important in responding to Miller (1985) and others who wish to deny the novelty of the new Christian right. The sense of grievance which led to the mobilization of the NCR is qualitatively different from early conservative Protestant political movements. The rejection of liberal cosmopolitanism is common to the fundamentalism of the 1920s and the 1970s but the replacement of world communism by secular humanism as the bogeyman represents an important change. Communism was 'out there' somewhere. Although fundamentalist leaders such as McIntire and Hargis (Forster and Epstein 1964) followed Senator McCarthy in seeing signs of communist infiltration in the heart of American institutions, communism was always a somewhat remote threat. The most recent wave of fundamentalist politicization is a response to changes which need very little social construction to appear to be just beyond the glow of the camp-fire.

Increasing secularity and liberalism could have been ignored by fundamentalists so long as they were permitted the social space in which to create and maintain their separate social institutions. Unfortunately for them, it is in the nature of modern industrial societies to reduce that social space.

There is nothing new about such boundary disputes. There have been many previous contests between particular religious minorities and the state. The Mormons were forced to abandon polygamy. Jehovah's Witnesses and Christian Scientists have come into conflict with the state over their refusal to permit such medical interventions as blood transfusions. In most states some sort of accommodation has been achieved, with religious exceptions to general laws being permitted. But what distinguishes the Mormons, the Witnesses, and the Christian Scientists is that they were despised minorities which were pleased to be tolerated and which had no great imperial ambitions. What makes the recent church–state conflicts serious is that they involve a very large minority which has imperial ambitions. Not surprisingly, given that their beliefs and values, language and thought once dominated very large parts of America, conservative Protestants want to see themselves as a moral *majority*.

To summarize, the people whom Jerry Falwell represents have not grown dramatically in numbers in the last fifteen years, although their ability to utilize new technology has given them a raised public profile. Liberals had simply forgotten that large numbers of people did not share their beliefs and values. The cosmopolitans and intellectuals who supervised the media and ran the bureaucracies of the major denominations had concentrated on the struggles for the rights of women and blacks, on the student movement, and on the protests against the Vietnam war, and neglected the American conservative Protestant. It is not so much fundamentalism but *public awareness* of fundamentalism which has been born again. In so far as some element of growth was involved in the rise of the new Christian right, it was the growth of evangelical and fundamentalist self-confidence, part of which came as an incidental feature of the rise of the sunbelt.

Similarly, the increased politicization of fundamentalism should not be seen as a reaction if by that one implies that it is fundamentalists who have changed markedly. In so far as there is a reactive element, it is not in the beliefs and values of fundamentalism but in the subculture's recognition that, to hold what it had and to avoid losing more, it must actively resist. If one is looking for a single word to describe the rise of the new Christian right, then 'reassertion' would be appropriate.

FIGHTING BACK

Fundamentalists are moved to fight back, either by changing the content of liberal and cosmopolitan culture or, more typically, by resisting the incursions of that culture and demanding the right to social space. I have tried to show that such resistance has a series of ironic consequences. To have any chance of success, it must concede many of the things which fundamentalists wish to preserve. When the concessions are not made, the NCR's claim to being a serious political movement is undermined. Given the smallness of the conservative Protestant population and its regional concentration, realistic political action requires pragmatism and accommodation. To have any hope of maintaining the social practices which they believe their religion requires, fundamentalists must compromise what is distinctive about that religion. In the world-view which creates the particular reasons conservative Protestants have for resisting modernism, Catholics and Jews are not Christians, and Mormonism is a dangerous cult. But legislative and electoral success requires that fundamentalists work in alliance with such groups and with secular conservatives.

The final irony of the position of contemporary fundamentalists is revealed in their attitude to minority rights. Fundamentalists object to the language of group rights, firstly because their religious and political ideologies are constructed around individualism, and secondly because the groups—racial minorities, feminists, and homosexuals—which have so far deployed the notion of group rights represent causes which fundamentalists have to date opposed. Nevertheless, the new Christian right has been most successful in the public arena when it has presented its own cause as being that of an oppressed and hard-done-by minority. As yet this is a rhetoric that only occasionally appears in fundamentalists' complaints about the neglect of their values and in the strategic thinking of NCR lobbyists and lawyers. But it is easy to imagine it taking hold and serving as a vehicle for coming to terms with modern America. The regionalism expressed in 'states' rights' will become the pluralism of minority rights and with it will come the end of any dream of a Christian empire.

THE DEFEAT OF THE NCR

It is perhaps premature to explain the failure of a social movement which has not yet died, but enough of the signs are already there—in fact, have always been there for those who wished to see them—for us to attempt an outline explanation.

The first point to make is that the potential support base for the NCR was always smaller than many commentators noticed. Firstly, the movement has to date failed to attract the support of any significant number of conservative Catholics, Jews, Mormons and others. But even if one confines the examination to conservative Protestants, it is clear that not all the members of those denominations which can sensibly be described as 'conservative Protestant' share the religious or political beliefs of Jerry Falwell. Commentators who notice only the growth of conservative Protestantism relative to the mainstream denominations are liable to see it mistakenly as an homogeneous or coherent base for the NCR.

Consider the Southern Baptist Convention, the largest grouping of conservative Protestants. Much has recently been made of the capture of a number of key organizational positions in the

Convention by active supporters of the NCR.[3] One might suppose that the entire membership of the SBC can be counted as present or future NCR supporters, and journalists do make this mistake. Actually it is likely that more than half the SBC affiliated congregations would stand apart from the NCR, either because they are sufficiently liberal to disagree with NCR positions or because, while themselves holding to such positions, they accept the right of others to disagree. A survey of the attitudes towards the Moral Majority of 431 SBC pastors showed that fewer than half were either members or supporters of the movement. Almost all the others described themselves as opponents (Guth 1983: 120).

The rift between 'orthodox' conservative Protestants and those who are becoming slightly more liberal is not the only important division. Supporters of the NCR are also criticized from the separatist fundamentalist position associated with Bob Jones University and its graduates. In one of a series of pamphlets entitled *Fundamental Issues in the 80's*, John Ashbrook concludes his critique of Falwell's *The Fundamentalist Phenomenon* by arguing that Falwell has changed camps. Falwell's concern to promote social issues has led him to abandon his separation from apostasy: the touchstone of fundamentalism. Another pamphlet in the series is called *Enforced Morality Does Not Produce Revival.* Bob Jones III has clearly expressed the view of this element of conservative Protestantism:

> The aim of the Moral Majority is to join Catholics, Jews, Protestants of every stripe, Mormons etc., in a common religious cause. Christians can fight on the battlefield alongside these people, can vote with them for a common candidate, but they cannot be unequally yoked with them in a religious army or organization. Morality is a matter of religion: a man's morality is based upon his religious beliefs. . . . Alliances we would avoid at the local level are not made acceptable or less ecumenical due to the national level on which they operate. . . . A close, analytical, biblical look at the Moral Majority . . . reveals a movement that holds more potential for hastening the church of Antichrist and building the ecumenical church than anything to come down the pike in a long time, including the charismatic movement. (Bob Jones III 1980: 1–3)

That there is an element of truth in the Jones assessment is suggested by Falwell's response to the crisis in the *PTL* ministry which

[3] On recent shifts of power in the SBC see Ammerman (1985) and *Time* (29 June 1987: 34).

followed Bakker's disgrace. Falwell is a traditional fundamentalist: *PTL* is a pentecostal ministry. In theory, Falwell should have regarded *PTL* as the purveyor of an unbiblical deceit. Instead, he pledged to do all he could both to preserve the organization and to maintain its audience, and assured its supporters that it would continue to be pentecostal. He defended this compromise of fundamentalist orthodoxy on the ground that the collapse of the PTL network would be seen by liberals and secularists as a victory. In the face of such enemies, pan-Christian solidarity is more important than doctrinal rectitude. It seems that the pragmatic accommodation originally advanced for the Moral Majority but denied for the religious sphere has spread from the political to the religious.

Conservative Protestantism is internally divided by theology and by differences over the correct way to interpret the need for 'separation from apostasy'. There is also a third source of division which is, to some extent, related to this second point. Conservative Protestants are also divided on the issue of legislating righteousness. Their ambivalence about imposing their morality on others who do not share their views has its roots in a genuine commitment to democracy. At points I have talked about pragmatic accommodation as if it were something distasteful which the circumstances of the American polity has forced on the NCR. While there is a certain economy in presenting the story in that manner, if left unqualified it does an injustice to the conservative Protestant tradition. Conservative Protestantism contains within it a tension between the obligation for the saints to rule righteously (even if that means imposing righteousness on the unregenerate), and the equally strong commitment to every individual's ability to discern the will of God. Although the latter tendency is theocratic rather than democratic, in practice it results in the same thing, and it is certainly the case that conservative Protestants have historically played a major part in the promotion of democracy and bourgeois individualism. It is the tragedy (in the strict sense of the word) of conservative Protestantism that one of its most valued consequences also undermines the conditions for its survival as the key source of values for a whole society rather than as the partial world-view of a self-selecting minority of saints. This is the moral to be drawn from the many surveys which show ambivalence among conservative Protestants about movements such as the

Moral Majority. On the one hand, conservative Protestants want to see the world 'returned to biblical standards'. On the other, their theology, ecclesiology, and history give them a strongly felt commitment to freedom of choice.

Some conservative Protestants choose pietistic retreat from the world on its own intrinsic merits as an alternative to imposing their righteousness on the unregenerate. For others it has been a sensible reaction to the failure of their more activist phases. The Reformed Presbyterians are a good example. In the early days of the Reformation in Scotland the Reformed Presbyterians (or Covenanters) were the most radical 'impositionists'. When they failed to win over the majority of the Church, they maintained their rhetoric of the saints and the civil magistrate working in a Godly harmony, while effectively retreating from the world. The American conservative Protestant tradition displays frequent alternation between periods of active involvement and retreat. Even when the activist element has been dominant there have been those who decry social and political involvement as a diversion, a waste of the energy which should be directed to the primary task of saving souls. It needs very little by way of disillusionment with the active mode to swing the pendulum back to pietistic retreat. Not being complete retreatists, the followers of the Bob Jones position encourage Christians to be politically active but insist on maintaining such a clear separation from apostasy that concerted political action is almost impossible.

One Moral Majoritarian described fundamentalists as a 'disciplined charging army' (Fitzgerald 1981). A political scientist called them 'an army that meets every Sunday' (Buell 1983). It would be more accurate to see them as a motley crew of half-hearted volunteers being pressed into service just when the crops need planting, torn between joining battle with the enemy and returning to tend their farms.

ECONOMIC AND SOCIAL CLEAVAGES

Further sources of internal fragmentation derive from socio-economic differences among conservative Protestants. Without wishing to countenance the a priori assumption of many social scientists that religious values are secondary to more concrete economic and social characteristics such as wealth and status, it is worth remembering that religious values compete with more mundane interests in the decision-making of conservative Protestants. While it is certainly the case that conservative Protestants differ from their more liberal brethren in giving a higher priority to 'biblical' positions, their interpretation of biblical injunctions and their willingness to act on such interpretations differ. In the period leading up to the Civil War, people with the same theology and ecclesiology developed fundamentally different attitudes towards the slavery issue, and conservative denominations divided into northern and southern branches. There may be no single secular issue likely to produce such an emotive or clear division within contemporary conservative Protestantism, but regional and status differences remain fissures which prevent conservative Protestants thinking and acting as a coherent body. I have already mentioned the divison between what, for brevity, I will call North and South on economic liberalism. Northern conservative Protestants are much more comfortable with doctrines of *laissez-faire* than are southerners, who profit considerably from the government spending.

Another division concerns race. Although most conservative Protestants are less likely than liberals to be supporters of racial integration, they are divided between those who are willing to support segregationist independent schools and social action to maintain residential segregation, and those who openly reject segregationist views and actions.

Finally, there is the neglected but important issue of priorities. The large number of surveys mentioned in this work (and many others) show that religion is important as a source of political images and decisions. We know that religion matters; what is more difficult to know is its place in the hierarchy of concerns on which any person draws to make a political decision. It is likely that some of the differences in the conclusions of surveys which have tried to measure the importance of theology or denominational affiliation are a result of individuals reordering their priorities. Some reordering may be caused by the questions asked in surveys. The very fact of being asked about one thing rather than another may produce a temporary reassessment of concerns. But it also seems likely that issues vary in salience depending on what is going on in the immediate world of the respondent. We can imagine conservative Protestants arranged on an axis of orthodoxy and suppose that those at the orthodox end consistently give a higher priority to their religious beliefs and values than do those at the liberal end. But at any point on the axis, the relative importance of religiously rooted values will vary with the events which impinge on their lives. While we may assume that values have a certain enduring quality, it also seems clear that their salience varies in response to 'agenda-setting' events in the world. If abortion is a topical issue, something which is frequently addressed and debated in the media and which features in elections, it will remain high on the list of priorities for those people who have strong views on the subject. When abortion slides down the public agenda, it will also seem less pressing for many of those who have strong views, not because their views have changed but because there is pressure and opportunity to address other issues.

Clearly a section of enthusiasts will strive to keep the spotlight of public attention focused on their concerns, but they have to compete for public attention with enthusiasts for other causes. In so far as any particular group of moral entrepreneurs has only a limited ability to set political agendas, there will be long periods when the issues around which conservative Protestants can unite are not very high on the lists of priorities even of conservative Protestants. To take an example from early 1987, the scandal over the sale of arms to Iran and the diversion of funds to the Nicaraguan Contras caused considerable media and public attention to focus on the honesty and competence of President Reagan and his officials. There was so much interest in 'Iranscam' and so many powerful forces working to keep attention focused on the general competence of the Reagan Administration that NCR lobbyists found themselves and their concerns marginalized. In very simple and practical terms, they could no longer get important people on the telephone. Even their sympathizers were engaged in other matters.

The same point can be made about many of the local elections in November 1986. In many areas, NCR activists found it impossible to make 'pro-family' concerns topics of interested debate. Instead most people, even many conservative Protestants, were interested in the economy. Even if the absence of NCR issues on the local political agenda did not cause potential supporters to give a higher priority to economic interests or foreign policy concerns, it almost certainly caused many to hold back from the debate and from voting. This problem is well appreciated by new Christian right activists. Much of their time is taken up, not with mobilizing support for a certain position on an issue, but with trying to make something which concerns them into a public issue.

LIBERAL REACTION

The exaggerations of the cohesion and commitment of the supporters of the NCR which were characteristic of many early responses to the NCR were often accompanied by almost total neglect of the power and influence of various liberal groups. The NCR was sometimes described as if its opposition had already been eradicated. In particular, a point of chronology which has considerable implications for the future was neglected. The organizational infrastructure of the new right and the NCR were described in terms which suggested that their sophisticated fund-raising and opinion formation techniques were novel. Far from this being the case, until the late 1970s the majority of ideological PACs were liberal. Had the accounts of the mobilization of the NCR made more of the fact that many of the movement's techniques were borrowed (even if, like direct mailing, they were considerably improved in the borrowing) from liberal causes, it might have been obvious earlier that the NCR would not have a clear run.

Although as a very general proposition liberals are more fragmented than conservative Protestants, it is possible for them to form effective organizations and to campaign for their goals when they feel sufficiently moved to act. With hindsight it is easy to see that much of the success of the NCR was due to the element of surprise. The resolution of the Texas textbook controversy, the defeat of the Arkansas and Louisiana 'equal time for creation science' bills, and the rejection of Judge Robert Bork show how effective liberals can be once they realize that they can no longer assume their views will naturally triumph but must actively promote them. In the field of electioneering, the right-wing steamroller which removed McGovern, Bayh, Church, and other liberals appeared unstoppable but those liberals who confronted negative campaigns head-on, rather than ignoring them as beneath contempt, won and, in some cases, improved their vote.

On the national stage, People for the American Way has become a highly efficient counter to the NCR. Having raised considerable sums of money, PAW is now in a position to counter the NCR with the same sophisticated technology, as an example will show. In September 1986, Pat Robertson staged a huge rally to test the waters for a presidential election campaign. As many as 216 conference halls throughout the country were booked and satellite time was leased to telecast the rally direct to around 200,000 potential supporters. The aim of this expensive operation was to bypass the conventional news media, which Robertson supposed would be unsympathetic to his ambitions. To counter this initiative, PAW booked a second satellite and offered free to any television station in the country that wanted it, a short programme of film clips of Robertson saying the sorts of things which would have been acceptable to the narrower audiences for which they were originally intended but which were potentially damaging to his new image as aspiring presidential candidate. In many of the major markets television stations took advantage of the free PAW material and inserted parts of it into their news reports of the Robertson rally. Although stations which were already sympathetic to Robertson's views ignored the PAW feed, many non-committed stations, which might otherwise have given Robertson some excellent publicity (his intention), presented a less flattering picture of the candidate. What had been designed by Robertson's organization as a good publicity return on a considerable investment was countered by an action which had the three qualities hitherto primarily associated with the NCR: imagination, good organization, and heavy funding.

Initially begun with a short anti-NCR television commercial, PAW has grown in six years to a membership of around 250,000 and a budget of $7.6 million dollars. A weekly radio commentary on church and state issues is provided free to radio stations around the country. Staff writers produce weekly opinion pieces which are offered free to local newspapers, many of which are only too happy to have the free, professionally produced, copy. A list of well-known public figures who will speak against the NCR is maintained so that people with appropriate expertise can be rapidly mobilized to present a liberal view on whatever is topical.

As Chapter 6 suggested, a major problem for NCR leaders is the need to alternate between two rhetorics: one designed for the faithful audience of fundamentalists; the other directed at non-fundamentalist potential allies. In his 1984 re-election campaign,

Representative Mark Siljander, an adviser to Christian Voice, sent a letter to some 400 conservative pastors in his district urging them to help defeat Jewish Michigan Democrat Howard Volpe and 'send another Christian to Congress'. Unfortunately, one of the conservative pastors had moved on and been replaced by a more liberal man who made sure that PAW and the press saw this evidence of religious particularism. Whereas in the first years of the NCR such disclosures were haphazard, PAW now maintains a staff of researchers who assiduously monitor the writings and sermons of Falwell, Robertson, Swaggart, and others so that the views they would prefer to express only to the inner circle of the faithful are given wide exposure.

In addition to closely monitoring NCR leaders in order to undermine their painstakingly calculated 'presentations of self' (to use Erving Goffman's term), liberals have themselves become adept at skilful self-presentation. Alert to the damaging consequences of being seen as avowed secularists, PAW and other similar organizations have stressed their support from Christians with impeccable conservative theological credentials. Three of the most prominent spokesmen in the campaign against the NCR are John Buchanan and James Dunn, both Baptist ministers, and Chuck Bergstrom, a respected conservative Lutheran clergyman. Such men have been able to argue that, far from promoting the Protestant cause, the NCR is debasing true religion. Instead of criticizing school prayer on secularist grounds, they can argue that public prayer in school trivializes their faith. With obvious conviction they can rest their case for the separation of church and state on the sentiments voiced by Justice Black in the majority opinion on the landmark *Engel* v. *Vitale* case:

> It is neither sacrilegious nor antireligious to say that each separate government in this country should stay out of the business of writing or sanctioning school prayer and leave that purely religious function to the people themselves and to those [*to whom*] the people choose to look for religious guidance. (in Abrahams 1983: 90)

THE SPECTATORS

Most Americans are neither committed liberals nor committed conservatives. In so far as we can tell from the complex mass of survey data and other sources of information, there is considerable ambiguity, both in responses to NCR issues and in feelings about NCR leaders and organizations. In the end, a major determinant of the very limited success of the NCR will be the ability of fundamentalists to present themselves as a legitimate minority. To create the social space required to maintain their subcultures and sub-societies, conservative Protestants will have to be able to convince the general public and the social groups with power and influence that they have legitimate rights which are presently being infringed, and that such rights can be accepted without significant social instability. There are problems for both actors and audience with this performance.

For the conservative Protestant now claiming to be a member of a hard-done-by group, the problem is one of muting demands. The belief that the survival of America as a nation requires the predominance of conservative Protestant beliefs and values is still too strong for the new mask of moderation not to slip. Changing Moral Majority Inc. to the Liberty Federation does not stop fundamentalists dreaming of the righteous empire. It is almost asking too much of committed conservative Protestants to expect them, in Falwell's words, to 'coalesce with fellow Americans with whom they have theological disagreements for the purpose of effecting moral and social change' (1986: 1). However inelegantly Falwell may put it, conservative Protestants do not have theological disagreements with other people. They have the truth and other people are wrong. It is difficult for them now to deflate their self-image from that of a 'moral majority' to that of a minority which asks nothing more than the right to do what is right in its own eyes.

The problem from the point of view of the audience is the authenticity of this new character. That many white, heterosexual, male Americans accepted that blacks, homosexuals, and women had been relatively disadvantaged was an important element in the limited success these groups have enjoyed in presenting themselves as minorities deserving of first tolerance and then positive discrimination. Conservative Protestants once played a major part in setting the social and moral tone of the nation; they succeeded in making alcohol consumption illegal. They have often been in the forefront of campaigns to prevent the extension of toleration to other groups. They were anti-Catholic, anti-immigrant, anti-semitic, anti-black, anti-homosexual, and anti-feminist. Groups which have never enjoyed power have not had the opportunity to behave badly towards others. Hence they can appeal to fairness without having that appeal undermined by the record of their own previous actions. Conservative Protestants once enjoyed considerable power (and are still powerful in some regions). While their exercise of power was not consistently malign, neither was it uniformly benign. It is easy for uncommitted observers to be suspicious of the change of the title of Falwell's organization. While Falwell would always have denied imperialist and impositional ambitions, the word 'majority' always carried the sense that, even if conservative Christians were not a majority, they would behave as if they were by imposing their views on society as a whole. Liberty Federation is more consistent with the minority rights strategy but many spectators doubt the sincerity of the new posture.

To summarize, any balanced assessment of the NCR has to consider the internal weaknesses of the movement as well as its strengths. It must also consider the effectiveness of liberal opposition. I have suggested that the failure of the earlier grand designs for cultural reformation has led the NCR to experiment with the potentially more fruitful strategy of claiming social space for its beliefs, values, and practices on the grounds that it represents a legitimate minority which the modern secular state discriminates against. The difficulty with that rhetoric is that many other groups have good reason to remember the lack of generosity of conservative Protestants when they were in the ascendant.

RELIGIOUS PARTICULARISM AND MODERNITY: CONCLUDING THOUGHTS

Twenty years from now, scholars will be in a much better position to judge the impact of the new Christian right. This study has argued that the power and influence of the movement have been greatly exaggerated, by its enemies as much as by its friends. The NCR has failed to achieve any significant legislative success, it has failed in its main goal of re-Christianizing America, and there are few reasons to suppose that it will at some future time succeed. Thus far the movement's failure has been explained in terms of complex relationships between the interests of potential supporters of the NCR and the circumstances in which they must work. In concluding, I want to draw attention to a more abstract concern: the problem of modernity. Underlying the particular organizational and motivational problems already discussed is the fundamental difficulty that what has brought the NCR into being is so amorphous as to be barely identifiable while at the same time being irreversible: what troubles supporters of the NCR is modernity and it will not go away.

WHAT IS SECULAR HUMANISM?

It is not usually the job of the sociologist to correct people's apprehensions of the world. If people define situations as real, then they are, for them, real. As a dictum for explaining why people do what they do, this is excellent. But when we move from explaining action to explaining why those best-laid schemes 'gang aft agley', a contrast between the way people define situations and our understanding of them is useful.

Most of the leaders and supporters of the NCR suppose that their grievances are the result of a conspiracy by an identifiable group of secular humanists. They are wrong. The Humanist Manifestos are red herrings, the credos of a small handful of not especially influential intellectuals.[4] Godless America of the 1980s is no more the creation of secular humanists than the America of the 1950s was the creation of communists. Were major social changes the result of identifiable people acting deliberately, consciously, and in concert, there would be better reasons to suppose the NCR might succeed in reversing them. A more accurate understanding of the source of the changes that disturb the NCR allows us to see the impossibility of the mission. What is required for that understanding is some notion of the distinctiveness of the modern world.

Berger *et al.* reasonably define modernization as the 'institutional concomitants of technologically induced economic growth' (1973: 15). The most important of these are directly related to the economy and closely related to them are 'the political institutions associated with what we know as the modern state, particularly the institution of bureaucracy' (1973: 16). Many things could be said about the consequences of technology and the modernity that

[4] Even Hunter (1986: 6) cannot find many humanists on any narrow definition. Martin (1981: 234) says: 'There can't be many true secular humanists, since only 3 per cent of all Americans say that they do not believe in God, and only a tiny fraction of those belong to the American Humanist Association or to other oragnizations that might qualify as denominations of the "religion of secular humanism".'

accompanies it; this brief description will concentrate on professionalism and pluralism.

Taking the second point first, modernity brings with it a near-inconceivable expansion of the area of human life which is open to choice.[5] Pre-modern man lived in a world of *fate*. In a world of only limited technology, the one tool was accompanied by a belief in the one way:

> One employs this tool, for a particular purpose and no other. One dresses in this particular way and in no other. A traditional society is one in which the great part of human activity is governed by such clear-cut prescriptions. Whatever else may be the problems of a traditional society, ambivalence is not one of them. (Berger 1980: 12)

This is not to say that traditonal societies are static, they do change, but their institutions—their routinized patterns of action—are marked by a high degree of certainty and 'taken-for-grantedness'. In most areas of life, things are done this way, and have always been done this way, because 'that is how we do things'. Modernity *pluralizes*: 'Where there used to be one or two institutions, there are now fifty. . . . where there used to be one or two programs in a particular area of human life, there are now fifty' (Berger 1980: 15).

Pluralization brings the need for choice. In contrast to the world of fate inhabited by traditional man, in innumerable situations of everyday life modern man must choose, and the necessity of choice reaches into the areas of beliefs, values, and world-views. For the modern society as a whole, pluralization requires that the state become ever more universalistic. Increased social differentiation and migration make populations less homogeneous. The gradual expansion of economies and of the state makes variations in ethnos, in religion, in race, and in language ever more troublesome. In traditional economies people trade preferentially and particularistically. In the modern capitalist economy, production and distribution are universalized. Although the process is never complete, the tendency is for modernizing societies to treat ever-larger proportions of the people in 'the same way'. The expansion of citizenship rights sees the universalizing of the franchise, of property rights, and of welfare provisions. The expansion

[5] This section (indeed my general sociological perspective) draws heavily on the work of Peter L. Berger and his interpretation of Max Weber and Arnold Gehlen (Berger 1973, 1979, 1980; Berger *et al.* 1973; and Berger and Luckmann 1973).

of bureaucracy—the application of technological rationality to the processing of people—sees increasing slices of identity being reduced to files consisting only of data relevant to 'the business in hand'.

A modern democratic nation-state which contains a variety of religious, racial, and ethnic groups and which wishes to be regarded as legitimate by the bulk of its population has to push religious, racial, and ethnic particularism out of the public arena and into the private 'home' world of individuals and their families. Only in the home is there sufficient consensus to prevent strongly held views and social identities being sources of social conflict. Put simply, a major consequence of pluralization is privatization.

The NCR sees the state imposing a coherent ideology which it calls 'secular humanism'. This is profoundly mistaken. What is actually imposed (and that term already suggests an inappropriately directed and conscious cause) is not so often the alternative dogma but the *dogma of alternatives*.

The main exceptions to this relativism occur when powerful professional groups claim dominion over some area of knowledge or action. Then the state may impose a particular view, as the Supreme Court has tacitly permitted in refusing to see creation science as a legitimate alternative to naturalistic explanations of the origin of species. However, the state endorsement of knowledge which competes with fundamentalist views is not the sole source of grievance, and it may not be the most threatening. To offer a flippant but none the less useful analogy, more cars are destroyed by rust than by crashes. The large obstacle that one can see is the obstacle one can avoid. As the strength of religion in Poland or the Soviet Union demonstrates, the state's attempts to produce its own functional equivalent of religion inadvertently encourages a reaction and makes it possible to continue to believe that there can be one truth, one shared vision, one world-view. The *contents* of competing visions can be ignored, or considered and then rejected. What cannot so easily be ignored is the constant evidence that there are many alternative visions. In the early stages of pluralism, some of the alternatives can be dismissed by invidious stereotyping of the proponents of those views. Especially when the carriers are foreign or largely confined to a status group quite different from one's own, the alternative world-views can be neutralized: Catholicism is the creed of Rome and rebellion; unitarianism and humanism are the creeds of degenerate upper classes; enthusiastic pentecostalism is the faith of the lumpenproletariat. The problem with late pluralism (to coin a parallel to the Marxists' late capitalism) is that a combination of proliferation of alternatives and social mobility makes such sanitizing-by-stereotype increasingly difficult. When there is so much variation in and across all social strata, even the most successful techniques for 'cognitive insulation' fail to disguise the reality of choice.

A major task of NCR ideologues has been to present their situation as one of being persecuted. At the end of Chapter 3, the social construction of secular humanism was described in terms of the value it had in simplifying the many grievances into one identifiable, embodied enemy. That observation can now be extended by returning to the importance of the 'two alternatives' argument used by the advocates of creation science in the Arkansas case. In that trial, and in the Alabama charge that school books taught secular humanism, fundamentalists insisted that there only were two forms of knowledge: fundamentalism and everything else. Anything which was not openly supportive of fundamentalism must be critical of it. It was with the purpose of reinforcing and extending that claim that the plaintiffs in the Alabama case called, as an expert witness, sociologist James D. Hunter, who argued that secular humanism was the functional equivalent of a religion. In a more detailed presentation Hunter (1986) defines humanism quite narrowly and makes the point that humanism differs from other religions in the very limited degree of consensus and coherence it engenders amongst its adherents. This important qualification is missed by fundamentalists. I want to go much further than Hunter and argue that the elements of humanism are so loosely articulated (in the mechanical rather than the rhetorical sense) that even his more refined presentation is misleading. For reasons which need not concern us here, I reject the value of defining religion in terms of its functions. But leaving aside the question whether humanism is a 'functional equivalent' of religion, it should be clear that it does not have the consequences—in providing a *common* direction to people's lives and a *shared* world-view—of, say, fundamentalism or traditional Catholicism. If humanism is defined narrowly, its support is insignificant. If it is defined broadly, as it is in the discourse of the NCR, then it is not an identifiable ideology. Instead it is the aggregation of everything that fundamentalists do not like.

The social construction work of NCR ideologues is directed towards disguising that fact. The homogenizing of secular humanism and the postulation of an active group of secular humanists are useful to the NCR in arguing its minority rights case but as a tool of social analysis they are profoundly misleading.

If one turns back to the definition of secular humanism offered by a Texan NCR organization—Pro-Family Forum—given at the end of Chapter 3, one finds that the Forum objects to: (a) the questioning of fundamentalist Protestant beliefs; (b) the rejection of the possibility of a whole society sharing the same detailed moral values; (c) the tolerance of alternative sexual 'life-styles'; (d) feminism; (e) socialism; (f) government controls on business activity; and (g) dramatic medical interventions connected with birth and death. It may be possible to imagine a modern democratic society which rejected the last four. What is not possible is to imagine one which could satisfy the NCR's desire to remove the first three.

Supporters of the NCR see the rejection of their religious beliefs and their commitment to a moral orthodoxy as the work of secular humanists. Certainly secular humanists believe in the removal of religion from the public arena, in the tolerance of alternative life-styles, and in the extension of choice but the sociologist of modernity sees the secular humanist position as being little more than the intellectual endorsement of what has already come to pass. While some of the changes which the NCR lump under the secular humanist label have been hastened by liberal moral entrepreneurs, most are the *unintended* consequences of modernity. When even those who are conservative on economic and foreign policy matters wish to retain the right to pursue their own life-styles, the only circumstance under which the NCR could succeed is a return to cultural homogeneity. Nothing visible to the student of the empirical social world suggests that the internal cultural fragmentation of modern societies is about to be reversed. In his analysis of the present, the sociologist thus becomes a curious bedfellow of the Bob Jones University fundamentalist; the necessary pre-condition for the success of the NCR is a massive religious revival. Where I differ from Bob Jones III is that I see no reason to suppose such a revival likely.

Were the grievances of American fundamentalists the result of the actions of secular humanists, they could be removed by the power of fundamentalist numbers expressed through the ballot box. After all, conservative Protestants remain one of the largest cultural minorities in America, and America is, generally speaking, a democracy. But at least part of what bothers fundamentalists is the apparent tension between items of modern scientific and technical knowledge and parts of the conservative Protestant world-view. To concentrate on evolution, it may well be that a modern industrial economy can permit the survival of pre-scientific ideas in certain limited spheres. The ability to make missiles, launch space rockets, exploit natural resources, and competitively produce cars may not be threatened by the belief that the world is less than a million years old and was made by God in six days. However, despite the willingness of Justices Rehnquist and Scalia to leave the matter of the origins of species to the vote of state legislatures, it seems clear that the general tendency of modern societies to accord priority in debates about matters of scientific interest to those with good credentials represents some sort of functional imperative, something that could not be different without posing a major threat to the knowledge base of the society. If that is the case, fundamentalists are not going to win their arguments with scientists and technologists, despite the occasional minor victory.

Something similar could be said of other areas of concern to the NCR. While there has been increasing hostility to the power of the professions, it still remains the case that in all advanced industrial societies, professionally accredited occupational groups dominate particular spheres of activity. Even on matters such as education, or the civil rights of the unborn or the terminally ill, where technical considerations are obviously informed by moral judgements, the opinions of professionals carry far more weight than those of lay people, largely because it is in the very nature of the modern society to translate moral and ethical matters into technical considerations (Wilson 1982: 42–52). The basic assumptions that inform modern industrial production—that all complex objects and procedures can be reduced to repeatable acts and replaceable components; that nothing is more than the sum of its parts; that everything can be measured and calculated; that nothing is sacred and that everything can be improved; that increased efficiency is the main imperative—cannot be confined to the world of work. The formal rationality which dominates that sphere gradually invades all other areas of social action. There is not the space here to present this argument in sufficient detail to convince the sceptical but it is accepted by most sociologists (of varying ideological positions) that it is characteristic of the modern world to subordinate the moral to the technical and the lay to the professional.

My point is that the authority of professionals (especially natural scientists) is such that fundamentalists are unlikely to establish the principle that arguments such as that over the origins of the species should be settled by votes rather than by the consensus of accredited experts. Even in matters that are more commonly seen as moral and ethical rather than technical and professional, the tendency is to defer to the professionals.

But even if one does not accept these claims about the scientific and technical consequences of modernity, one cannot think away the consequence of pluralism. And, if, as I suggest, secular humanism is simply a convenient blanket term for the necessary consequences of pluralism, then clearly fundamentalists have no hope of attaining their goals because what offends them is nothing more or less than modernity itself.

UNIVERSALISM AND THE NCR AS A LEGITIMATE MINORITY

The awkward position of the NCR can now be fully described by bringing the above observations about the universalizing tendencies of modern societies together with the earlier discussion of the NCR's fall-back position of presenting itself as a disadvantaged minority.

Blacks, women, and homosexuals have built their claims by pointing to the failure of parts of the economy, the polity, and the social structure to live up to the rhetoric of universalism. Far from challenging modernity, they have appealed to its core values by identifying areas in which universal principles regarding economic and political rights have not been rigorously pursued. They have presented themselves as discriminated against by the failure of the state to prevent the continuation of particularistic practices in employment, political representation, and social valuation. The demands of these minorities are thus, in theory, demands which can be met by a modern industrial society simply giving more effort to its existing dynamics. The outlawing of racism and sexism can be seen as merely giving more substance to the universalizing tendency. Racist and sexist language, for example, can be eradicated by stripping the culture of certain particularistic features; by making it more bland.

The NCR's claims to the status of a legitimate minority seem quite different. The supporters of the NCR are not disadvantaged in terms of socio-economic status (or at least, in so far as they are, it is because of their class, region, levels of education, and other characteristics not specifically related to their shared religious culture). Where they are disadvantaged is in the status which the state is willing to accord their culture. This disadvantage cannot be remedied by an extension of the twin principles of universalism in the public sphere and the relegation of particularism to the private world. It is precisely these two principles which have produced most of the changes which offend supporters of the new Christian right. Thus although the shift from (a) aiming to re-Christianize America to (b) claiming only that their values, beliefs, and symbols be accorded due status in the public arena is a sensible change in strategy for new Christian rightists, an understanding of the most abstract features of modernity gives every reason to suppose that it is a strategy doomed to fail.

THE NCR AS A MODERN PHENOMENON

There is a tendency to see the NCR as a reactionary movement, an outburst of resurgent traditionalism. Certainly its proponents are fond of describing it in terms such as those of the title of one manifesto: *Back to Basics* (Pines 1982). In part this characterization is appropriate but it is important also to stress the extent to which the movement has itself accommodated to modernity. This accommodation is not just a matter of adjusting rhetoric so that the religion of Creationism becomes creation science and the virtues of fundamentalism are presented, not as divine injunctions, but as socially functional arrangements. It is also a matter of conceding crucial ground to the pluralism of the modern world by accepting the need to separate religious values and socio-moral positions so that alliances can be formed with advocates of competing religious values. Leaders of the NCR insist that they have not accepted the denominational attitude (in which truth is relativized so a number of apparently and previously competing visions can all be seen as being in some sense equally valid) but they have accepted another crucial element of modernity; they compartmentalize. They operate in a world of social action that has been divided into separate spheres with different values. In church, with their own people, in prayer meetings, they remain fundamentalist Protestants. But when pursuing the public agenda of socio-moral issues they operate with a quite different set of criteria. That is, they have conceded a major part of what the modern pluralistic society demands of religion: its restriction to the private home world. Although their behaviour in the public sphere is still informed by religious considerations, it is not dominated by them, and they have been diluted in order to attract maximum support from people who do not share the values and beliefs of conservative Protestantism.

The alternative to denominationalism is sectarianism: the continued insistence that what one has is *the* truth and that those who differ are simply wrong. To present the situation of religion in a modern society in the starkest possible terms, the choice is between sectarianism and denominationalism. Modernity constantly increases the costs of sectarianism. Those people who wish to maintain orthodox religious beliefs find themselves having to retreat further and further into either regional or socially constructed laagers. The NCR has tried to reduce the costs, both by seeking public support for its positions and by resisting the encroachments of the central secular state. But in trying to do those things, it has been forced to accept the denominational attitude. One can see this clearly in conservative Protestant reasoning about the possibility of a third party. Where religion exists in its 'church' form, it does not need to be represented by a political party because its presence is so all pervading; Catholicism in the Republic of Ireland is a case in point. Where it exists in a sectarian form, it produces a coherent confessional party; the Calvinist Anti-revolutionary party in nineteenth-century Holland is an example. American conservative Protestants realize a confessional party is not a possibility. Those who talk about a religious party at all recognize that it would have to be at least a Christian or even a Judaeo-Christian party. But most of them realize that even a Judaeo-Christian party would not work; any viable third party would have to be a secular party informed by traditional (i.e. religious) values. That is denominationalism and it is a long way down the road to the point where religion is hardly a factor at all, where religiosity appears only through political attitudes which reflect general class and status interests. The situation becomes one where the second words in the phrases 'conservative Protestant', 'conservative Catholic', and 'conservative Jew' become redundant.

Modernity does not challenge religion. Instead it subtly undermines it and corrodes it. Fundamentalists tacitly recognize this when they refuse to be impressed or comforted by the state's willingness to permit—to *tolerate*—Mormons, Witnesses, Christian Scientists, Rastafarians, Scientologists, Moonies, and any number of more exotic religions. Although few fundamentalists say it openly, some of them recognize that it is better to be persecuted than to be tolerated as (in the language of American forms) a 'religion of your preference'.

Twenty or so years ago many of the sociologists who endorsed the above picture of modernity supposed that secularization—the decline of religion—was an irreversible characteristic of modern societies. Recently the sociological orthodoxy seems to have been running in the opposite direction. Although I remain committed to a version of the secularization thesis, I do not expect religion to disappear completely or quickly. And in so far as it is the broad liberal denominations which are losing support fastest, one would expect traditional supernaturalist Protestantism to become relatively more popular and influential. There is thus nothing surprising about the appearance of the new Christian right. So long as there are sizeable numbers of conservative Protestants in America, there will be movement organizations which campaign and lobby on their behalf. There will continue to be skirmishes and boundary disputes. Precisely because the conflict is not between two groups of believers, but between the adherents to a coherent belief-system and modernity, it will always be difficult to judge accurately the outcome of any battle. It will depend more on counting the dead on both sides than on watching to see who marches victorious from the field. What this study has tried to do is to consider calmly what is known about the support-base, the actions, and the impact to date of the NCR in order to evaluate the likelihood of it succeeding in its ambitions. The conclusion is that the NCR will fail—in its present form already has failed—both to re-Christianize America and to prevent further displacement of the values which its supporters hold dear. And—the point made by fundamentalist critics of the NCR—the very limited successes enjoyed by the movement have been won at the cost of submitting to modernity and abandoning the ethos of orthodox separatism which has been characteristic of fundamentalism.

AFTERWORD

Since the Moral Majority was officially chartered in June 1979, Jerry Falwell has rightly been seen as the predominant figure in the new Christian right. On 4 November 1987 Falwell announced that he was resigning from the presidency of Moral Majority and the Liberty Federation: 'I will not be stumping for candidates again. I will never work for a candidate as I did for Ronald Reagan. I will not be lobbying for legislation personally' (*Independent* 5 November 1987). With the failure of his mission to restore the finances of the Bakkers' PTL gospel television corporation and declining income for his own *Old Time Gospel Hour*, Falwell had good reason to rededicate himself to his gospel ministry. He could take some consolation from Pat Robertson's campaign for the Republican presidential nomination. He could and did take credit for 'breaking the psychological barrier that religion and politics don't mix'. But no amount of brave face could alter the impression that he was leaving politics a disappointed man. Less than two years earlier, when announcing the renaming of Moral Majority, he had said:

> with the Liberty Federation, we will be advancing to another level of involvement. We will also be challenging many of our people to run for office at the local, state and national levels. . . . We now sincerely believe that it is possible to form a coalition of religious conservatives in this nation, including The Liberty Federation and scores of similar groups, which can bring 20 million voters to the polls nationally by 1988. This is our goal. (1986: 2)

The best assessment of the NCR's chance of attaining that goal in 1988 or at any other time in the future is that given by Falwell himself in his decision to abandon the leadership of a movement in which he had invested so much of his time and prestige.

BIBLIOGRAPHY

Aberle, D. (1966). *The Peyote Religion among the Navaho*. Aldine, Chicago.

Abraham, H. J. (1983). *The Judiciary: the Supreme Court in the governmental process*. Allyn and Bacon, Boston.

Ahlstrom, S. E. (1972). *A Religious History of the American People*. Yale University Press, New Haven.

Alexander, C. C. (1965). *The Ku Klux Klan in the South West*. University of Kentucky Press, Kentucky.

Ammerman, N. (1985). 'Organizational conflict in a divided denomination', *SBC Today*, December, 12–14.

Ashbrook, J. E. (1982). *An Analysis by a Fundamentalist of Falwell's The Fundamentalist Phenomenon*. International Committee for the Propagation and Defense of Biblical Fundamentalism, Greenville, SC.

Barrett, D. B. (1982). *World Christian Encyclopaedia: a comparative study of churches and religions in the modern world, A.D. 1900–2000*. Oxford University Press, Nairobi.

Bartley, N. V. (1970). *From Thurmond to Wallace: political tendencies in Georgia, 1948–1968*. Johns Hopkins University Press, Baltimore.

Bell, D. (1964). *The Radical Right*. Doubleday Anchor, New York.

Bellah, R. N. (1967). 'Civil religion in America'. *Daedalus*, Vol. 96, 1–21.

Berger, P. L. (1973). *The Social Reality of Religion*. Penguin, Harmondsworth, Middx.

—— (1979). *Facing up to Modernity: excursions in society, politics and religion*. Penguin, Harmondsworth, Middx.

—— (1980). *The Heretical Imperative: contemporary possibilities of religious affirmation*. Collins, London.

—— Berger, B., and Kellner, H. (1973). *The Homeless Mind: modernization and consciousness*. Penguin, Harmondsworth, Middx.

—— and Luckmann, T. (1973). *The Social Construction of Reality: a treatise in the sociology of knowledge*. Penguin, Harmondsworth, Middx.

Berthoff, R. (1971). *Unsettled People: social order and disorder in American history*. Harper and Row, New York.

Bestic, A. (1971). *Praise the Lord and Pass the Contribution*. Cassells, London.

Billings, W. (1980). *The Christian's Political Action Manual.* National Christian Action Committee, Washington, DC.

Billington, R. (1964). *The Protestant Crusade, 1800–1860: a study in the origins of American nativism.* Quadrangle Books, Chicago.

Bishop, G. F., Tuchfarber, A. J., and Oldendick, B. W. (1986). 'Options on fictitious issues: the presssure to answer survey questions', *Public Opinion Qtly.*, Vol. 50 (2), 240–50.

Black, A. W. (1986). 'The sociology of ecumenism: initial observations on the formation of the Uniting Church in Australia', in A. W. Black and P. E. Glasner (eds.), *Practice and Belief: studies in the sociology of Australian religion*, 86–107. George Allen and Unwin, Sydney.

Blocker T. J. and Riedesel, P. L. (1978). 'Can sociology find true happiness with subjective status inconsistency?', *Pacific Sociol. Rev.*, Vol. 21 (3), 275–88.

Blumenthal, S. (1984). 'The righteous empire', *New Republic*, Vol. 46, 18–24.

Blumer, H. (1967). *Symbolic Interactionism.* Prentice-Hall, Englewood Cliffs, NJ.

Bollier, D. (1982). *Liberty and Justice for Some: defending a free society from the radical right's holy war on democracy.* People for the American Way, Washington, DC.

Box, S. and Ford, J. (1969), 'Some questionable assumptions in the theory of status inconsistency', *Sociol. Rev.*, Vol. 17, 187–201.

Brady, D. W. and Hurley, P. A. (1985). 'The prospects for contemporary partisan re-alignment', *Pol. Stud.*, Vol. 33, 63–9.

Brady, D. W and Tedin, K. L. (1976). 'Ladies in pink: religion and political ideology in the anti-ERA movement', *Soc. Sci. Qtly.*, Vol. 56, 564–75.

Bromley, D. and Shupe, A. (1979). ' "Just a Few Years Seems Like a Lifetime": a role theory approach to participation in religious movements', *Research in Social Movements, Conflict and Change*, Vol. 2, 159–65.

Brownstein, R. and Easton, N. (1982). *Reagan's Ruling Class: portraits of the President's top 100 officials.* Presidential Accountability Group, Washington.

Bruce, S. (1983). *One Nation under God: observations on the new Christian right in America.* The Queen's University of Belfast, Belfast.

—— (1985). *No Pope of Rome: militant Protestantism in modern Scotland.* Mainstream, Edinburgh.

—— (1986a). *God Save Ulster!: the religion and politics of Paisleyism.* Clarendon Press, Oxford.

—— (1986b). 'Militants and the margins: British political Protestantism', *Sociol. Rev.*, Vol. 34 (3), 797–811.

—— (1987). 'Status and cultural defense: the case of the new Christian right', *Sociol. Focus*, Vol. 20 (3), 242–6.

—— and Wallis, R. (1983). 'Rescuing motives', *Brit. J. Sociol.*, Vol. 34, 61–71.

Buell, E. H. (1983). 'An army that meets every Sunday? Popular support for the Moral Majority in 1980'. Paper given at Midwest Pol. Sci. Assoc., Chicago, April.

—— and Sigelman, L. (1985). 'An army that meets every Sunday? Popular support for the Moral Majority in 1980', *Soc. Sci. Qtly.*, Vol. 66 (2), 426–34.

Butts, R. F. (1986). *Religion, Education and the First Amendment: the appeal to history*. People for the American Way, Washington, DC.

Canovan, M. (1981). *Populism*. Junction Books, London.

Carroll, J. W., Johnson, D., and Marty, M.E. (1979). *Religion in America Since 1950*. Harper and Row, San Francisco.

Carroll, P. N. and Noble, D. W. (1982). *The Free and the Unfree: a new history of the United States*. Penguin, Harmondsworth, Middx.

Cash, W. J. (1954). *The Mind of the South*. Vintage Books, New York.

Clabaugh, G. (1980). *Thunder on the Right: the Protestant fundamentalists*. Nelson-Hall, New York.

Clark, D. W. and Virts, P. H. (1985). 'Religious television audience: a new development in measuring audience size'. Paper given at the Center for the Scientific Study of Religion, Savannah, Georgia.

Clubb, J. M., Flanigan, W. H., and Zingale, N. H. (1980). *Partisan Realignment: voters, parties and government in American history*. Sage, Beverly Hills.

Congressional Quarterly. (1982). 'Dollar Politics', 3rd edn., Congressional Quarterly Inc., Washington, DC.

Conway, F. and Siegelman, J. (1982). *Holy Terror*. Doubleday, New York.

Coxon, A. P. M., Davies, P. M., and Jones, C. L. (1986). *Images of Social Stratification: occupational structures and class*. Sage, London.

Crawford, A. (1980). *Thunder on the Right: the new right and the politics of resentment*. Pantheon, New York.

Currie, E. and Skolnick, J. H. (1970). 'A critical note on conceptions of collective behavior', *Annals Amer. Academy*, Vol. 391, 46–55.

Danzig, D. (1962). 'The radical right and the rise of the fundamentalist minority', *Commentary*, Vol. 33 (4), 291–8.

Davis, L. J. (1980). 'Conservatism in America', *Harper's*, Vol. 261, 21–6.

Demerath III, N. J. and Williams, R. H. (1987). 'A mythical past and an uncertain future', in T. Robbins and R. Robertson (eds.), *Church–State Relations; tensions and transitions*, 77–91. Transaction, New Brunswick.

Denenberg, R. V. (1984). *Understanding American Politics*. Fontana, London.

Dewar, H. (1983). 'Violence split West Virginia county in '74', *Miami News*, 28 April.

Drew, E. (1981). 'Jesse Helms', *New Yorker*, Vol. 57 (22), 78–95.

D'Souza, D. (1986). 'Jerry Falwell is reaching millions and drawing fire', *Conservative Digest*, December, 5–12.

Duke, P. (ed.) (1986). *Beyond Reagan: the politics of upheaval.* Warner Books Inc., New York.

Eitzen, D. S. (1970a). 'Status inconsistency and Wallace supporters in a mid-western city', *Soc. Forces*, Vol. 44, 493–8.

—— (1970b), 'Social class, status inconsistency and political attitudes', *Soc. Sci. Qtly.*, Vol. 51, 602–9.

Elms, A. C. (1970). 'Those old ladies in tennis shoes are no nuttier than anyone else, it turns out. Pathology and politics', *Psychology Today*, February, 27–59.

Fairbanks, D. (1977). 'Religious forces and "morality" politics in the American states', *Western Pol. Qtly.*, Vol. 30 (3), 411–17.

Falwell, J. (1979). *Listen America!* Doubleday, New York.

—— (1986). Official statement on launch of The Liberty Federation. National Press Club, 3 January 1986.

—— and Towns, E. (1971). *Church Aflame.* Impact Books, Nashville.

Ferguson, T. and Rogers, J. (1986). 'The myth of America's turn to the right', *Atlantic Monthly*, May, 43–53.

Ferguson, T. W. (1987). 'What next for the conservative movement?', *The American Spectator*, Vol. 20 (1), 14–6.

Finer, S. E. (1982). *Comparative Government.* Penguin, Harmondsworth, Middx.

Fishman, B. (1984). *American Families: responding to the pro-family movement.* People for the American Way, Washington, DC.

Fitzgerald, F. (1981). 'A disciplined charging army', *New Yorker*, 18 May, 53–97.

—— (1986). *Cities on a Hill: a journey through contemporary American cultures.* Picador, London.

Forster, A. and Epstein, B. R. (1964). *Danger on the Right: the attitudes, personnel and influence of the radical right and extreme conservatives.* Random House, New York.

Fumento, M. (1987). 'Some dare to call them . . . robber barons', *National Review*, 13 March, 32–8.

Furguson, E. B. (1986). *Hard Right: the rise of Jesse Helms.* W. W. Norton, New York.

—— (1987). 'Ambassador Helms', *Common Cause Magazine*, March/April, 16–21.

Gallup Organization. (1985). *Religion in America, 50 Years: 1935–1985*, Gallup, Princeton, NJ.

Galtung, J. (1964). 'A structural theory of aggression', *J. Peace Research*, Vol. 1, 36–54.

Gasper, L. (1963). *The Fundamentalist Movement*. Mouton, The Hague.

Gaustad, E. S. (1973). *Dissent in America*. Chicago University Press, Chicago, Ill.

Gerlach, L. P. and Hine, V. H. (1968). 'Five factors crucial to the growth and spread of a modern religious movement', *J. Sci. Stud. Rel.*, Vol. 7, 23–40.

Gibbs, J. P. and Martin, W. J. (1958). 'A theory of status integration and its relationship to suicide', *Amer. Sociol. Rev.*, Vol. 23, 140–47.

Gilkey, L. (1985). *Creationism on Trial: evolution and God at Little Rock*. Winston Press, Minneapolis.

Glenn, C. (1982). *Enforced Morality Does Not Produce Revival*. International Committee for the Propagation and Defense of Biblical Fundamentalism, Greenville, SC.

Godwin, R. S. (1983). 'Pro-lifers: a needed change in strategy', *Moral Majority Report*. January, 4–5.

Goffman, E. (1970). *Stigma: notes on the management of spoiled identity*. Penguin, Harmondsworth, Middx.

Goldstein, C. (1985). 'What Ronald Reagan needs to know about Armageddon', *Liberty*, Vol. 80 (6), 2–6.

Grant, A. R. (1979). *The American Political Process*. Heinemann, London.

Greene, J. (1981). 'The astonishing wrongs of the new moral right', *Playboy*, January, 117–18, 248–62.

Gurr, T. (1970). *Why Men Rebel*. Princeton University Press, Princeton, NJ.

Gusfield, J. (1963). *Symbolic Crusade: status politics and the American temperance movement*. University of Illinois Press, Urbana.

Guth, J. (1983). 'Southern Baptist clergy: vanguard of the new Christian right', in R. C. Liebman and R. Wuthnow (eds.), *The New Christian Right*, 118–31. Aldine, Chicago.

—— (1984). 'The politics of preachers: Southern Baptist ministers and Christian right activism', in D. Bromley and A. Shupe (eds.), *New Christian Politics*, 235–50. Mercer University Press, Macon, Ga.

Hadden, J. K. and Swann, C. E. (1981). 'Prime-time Preachers: the rising power of televangelism. Addison-Wesley, Reading, Mass.

Hammond, P. E. (1983). 'The courts and secular humanism', in T. Robbins and R. Robertson (eds.), *Church–State Relations; tensions and transitions*, 91–103. Transaction, New Brunswick.

Hamnett, I. (1973). 'Sociology of religion and the sociology of error', *Religion*, Vol. 3 (1), 1–12.

Harper, C. L. and Leicht, K. (1984). 'Explaining the new religious right: status politics and beyond', in D. Bromley and A. Shupe (eds.), *New Christian Politics*, 101–11. Mercer University Press, Macon, Ga.

Himmelstein, J. I. and McRae. J. A. Jun. (1984). 'Social conservatism, new republicans and the 1980 election', *Public Opinion Qtly.*, Vol. 48, 592–605.

Hofstadter, R. (1967). *The Paranoid Style in American Politics and other essays*. Vintage, New York.

Hohl, A. (1982). 'By one vote', *Moral Majority Report*, October, 12.

Hood, R. W. Jun. and Morris, R. J. (1985). 'Boundary maintenance, sociopolitical views, and presidential preference among high and low fundamentalists', *Rev. Rel. Res.*, Vol. 27, 134–45.

Howe, G. N. (1981). 'The political economy of American religion: an essay in cultural history, in S. G. McNall (ed.), *Political Economy: a critique of American society*, 110–35. Scott, Foreman and Co., Boston.

Hunt, L. L. and Cushing, R. G. (1970). 'Status discrepancy, interpersonal attachment and right-wing extremism', *Soc. Sci. Qtly.*, Vol. 51, 587–601.

Hunter, J. D. (1983). *American Evangelicalism: conservative religion and the quandary of modernity*. Rutgers University Press, New Brunswick.

—— (1986). 'Humanism and social theory: is secular humanism a religion?' Unpublished paper.

—— (1987). *Evangelicalism: the coming generation*. University of Chicago Press, Chicago, Ill.

Hutcheson, J. D. and Taylor, G. A. (1973). 'Religious variables, political system characteristics and policy outputs in the American states', *Amer. J. Pol. Sci.*, Vol. 17, 414–27.

Jackson, E. F. (1962). 'Status inconsistency and symptoms of stress', *Amer. Sociol. Rev.*, Vol. 27, 469–77.

—— and Curtis, R. F. (1972). 'Effects of vertical mobility and status inconsistency: a body of negative evidence', *Amer. Sociol. Rev.*, Vol. 37, 701–13.

Janovitz, M. (1978). *The Last Half-century: societal change and politics in America*. University of Chicago Press, Chicago, Ill.

Johnson, G. B. (1923). 'A sociological interpretation of the new Ku Klux movement', *Soc. Forces*, Vol. 1, 440–5.

Johnson, S. D. and Tamney, J. B. (1982). 'The Christian right and the 1980 presidential election', *J. Sci. Stud. Rel.*, Vol. 21 (2), 123–31.

—— (1985). 'The Christian right and the 1984 presidential election', *Rev. Rel. Res.*, Vol. 27 (2), 124–33.

Jones, M. A. (1960). *American Immigration*. University of Chicago Press, Chicago, Ill.

Jones, B. III. (1980). *The Moral Majority*. Bob Jones University Press, Greenville, SC.

Jorstad, E. (1970). *The Politics of Doomsday: fundamentalists of the far right*. Abingdon Press, Nashville.

King, L. L. (1966). 'Bob Jones University: the buckle on the Bible belt', *Harper's*, June, 51–8.

Koenig, T. and Boyce, T. (1985). 'Corporate financing of the Christian right', *Humanity and Society*, Vol. 9, 13–28.

Landsberger, H. (1969). *Latin American Peasant Movements*. Cornell University Press, Ithaca, NY.

Larson, J. R. (1982). 'The new right: its agenda for education', *American Secondary Education*, Vol. 11 (4), 2–5.

Latus, M. A. (1983). 'The politics of ideological and religious PACs', in R. C. Liebman and R. Wuthnow (eds.), *The New Christian Right*, 75–99. Aldine, Chicago.

—— (1984). 'Mobilizing Christians for political action: campaigning with God on your side', in D. Bromley and A. Shupe (eds.), *New Christian politics*, 251–67. Mercer University Press, Macon, Ga.

Lawrence, D. G. and Fleister, R. (1987). 'Puzzles and confusion: political re-alignment in the 1980s', *Pol. Sci. Qtly.*, Vol. 102 (1), 79–92.

Lenski, G. (1954). 'Status crystallization: a non-vertical dimension of social status', *Amer. Sociol. Rev.* Vol. 19, 405–13.

Liebman, R. C. (1983a). 'Mobilizing the Moral Majority', in R. C. Liebman and R. Wuthnow (eds.), *The New Christian Right*, 50–73. Aldine, Chicago.

—— (1983b). 'The making of the new Christian right', in R. C. Liebman and R. Wuthnow (eds.), *The New Christian Right*, 229–38. Aldine, Chicago.

Lienesch, M. (1982). 'Rightwing religion: Christian conservatism as a political movement', *Pol. Sci. Qtly.*, Vol. 37 (3), 403–25.

Lipset, S. M. (1964). 'Religion and politics in the American past and present', in R. Lee and M. Marty (eds.), *Religion and Social Conflict*,, 69–126. Oxford University Press, New York.

—— (1965). 'An anatomy of the Klan', *Commentary*, Vol. 150, October, 74–83.

—— (1969). *Revolution and Counter-Revolution*. Heinemann, London.

—— (1978). *Political Man*. Heinemann, London.

—— and Raab, E. (1978). *The Politics of Unreason: right-wing extremism in America, 1790–1977*. University of Chicago Press, Chicago Ill.

—— (1981). 'Evangelicals and the elections'. *Commentary*, Vol. 71 (3), 25–31.

—— Trow. M., and Coleman, J. (1956). *Union Democracy*. Doubleday Anchor, New York.

Lofland, J. (1985). *Protest: studies in collective behavior and social movements*. Transaction Books, New Brunswick.

Lorentzen, L. J. (1980). 'Evangelical life style concerns expressed in political action', *Sociol. Anal.*, Vol. 41, 144–54

Louks, E. H. (1936). *The Ku Klux Klan in Pennsylvania: a study in nativism*. Telegraph Press, New York.

Lyons, G. (1982). 'Repealing the enlightenment', *Harper's*, April, 38–78.

McCarthy, J. D. and Zald, M. (1973). *The Trend of Social Movements in America: professionalization and resource mobilization*. General Learning Press, Moristown, NJ.

—— —— (1977). 'Resource mobilization and social movements: a partial theory', *Amer. J. Sociol.*, Vol. 82 (6), 1212–41.

McGuigan, P. B. (1986a). 'Education Secretary William J. Bennett goes to the people', *Conservative Digest*, February, 47–58.

—— (1986b). 'Attorney General Edwin Meese III talks straight', *Conservative Digest*, April, 5–14.

—— (1986c). 'The religious and political values of Dr Pat Robertson', *Conservative Digest*, December, 31–8.

McIntire, T. (1979). *The Fearbrokers*. Beacon Press, Boston.

McLeish, J. (1969). *Evangelical Religion and Popular Education: a modern interpretation*. Methuen, London.

Manion, M. (1986). 'The impact of state aid on sectarian higher education: the case of New York state', *Rev. Politics*, Vol. 48 (2), 264–88.

Marsden, G. (1980). *Fundamentalism and American Culture: the shaping of twentieth-century evangelicalism*. Oxford University Press, London.

Martin W. (1981). 'God's angry man', *Texas Monthly*, April, 153–235.

Marty, M. E. (1969). *The Modern Schism: three paths to the secular*. Harper and Row, New York.

—— (1970). *Righteous Empire: the Protestant experience in America*. The Dial Press, New York.

Marx, G. T. and Wood, J. (1975). 'Strands of theory and research in collective behavior', *Annual Rev. Sociology*, Vol. 1, 363–428.

Marx, J. H. and Holzner, B. (1977). 'The social construction of strain and ideological models of grievance in contemporary movements', *Pacific Sociol. Rev.*, Vol. 20, 411–13.

Melton, J. G. (1978). *The Encyclopaedia of American Religions*. 2 Vols. McGrath Publishing, Wilmington, NC.

Menendez, A. J. (1977). *Religion at the polls*. Westminster Press, Philadelphia.

Michels, R. (1959). *Political Parties: a sociological study of the oligarchical tendencies of modern democracy*. Dover, New York.

Miller, A. H. and Wattenberg, M. P. (1984). 'Politics from the pulpit: religiosity and the 1980 elections', *Public Opinion Qtly.*, Vol. 48, 302–17.

Miller, W. E. (1985). 'The new Christian right and fundamentalist discontent: the politics of lifestyle concern hypothesis revisited'. *Sociol. Focus*, Vol. 18, 325–36.

Moffat, J. M. (1963). *The Ku Klux Klan: a study of the American mind.* Russell and Russell, New York.

Mueller, C. (1983). 'In search of a constituency for the "New Religious Right" '. *Public Opinion Qtly.*, Vol. 47, 213–29.

National Coalition Against Censorship. (1982). 'In High Court book-banning test, it's Pico 5, Island Trees School Board 4', *Censorship News*, October, 1.

Neely, R. (1981). *How Courts Govern America.* Yale University Press, New Haven.

Niebuhr, H. R. (1962). *The Social Sources of Denominationalism.* Meridian Books, New York.

Oberschall, A. (1973). *Social Conflict and Social Movements.* Prentice-Hall, Englewood Cliffs, NJ.

Olson, M. E. and Tully, J. C. (1972). 'Socio-economic-ethnic status, status inconsistency and preference for political change'. *Amer. Sociol. Rev.*, Vol. 37, 560–74.

Ornstein, N. J. (1986). 'The party's over', *State Legislatures*, November/December, 15–17.

Page, A. L. and Clelland, D. A. (1978). 'The Kanawha county textbook controversy: a study in the politics of life style concern', *Social Forces*, Vol. 57, 265–81.

Parker, B. (1979). 'Meet the textbook crusaders', *American School Board Journal*, June, 21–8.

—— (1983). *As Texas Goes, So Goes the Nation: a report on textbook selection in Texas.* People for the American Way, Washington, DC.

Parker, D. A. (1979). 'Status inconsistency and drinking behavior', *Pacific Sociol. Rev.*, Vol. 22 (1), 77–95.

Parsons, T. (1969). *Politics and Social Structure.* Free Press, New York.

Patel, K., Pilant D., and Rose, G. (1982). 'Born-again Christians in the Bible belt: a study in religion, politics and ideology', *Amer. Pol. Qtly.*, Vol. 10 (2), 255–72.

Peele, G. (1984). *Revival and Reaction: the right in contemporary America.* Clarendon Press, Oxford.

People for the American Way (1985). *Attacks on the Freedom to Learn: a 1983–84 report.* People for the American Way, Washington, DC.

—— (1986). *Press Clips.* People for the American Way, Washington, DC.

—— (1987). *Attacks on the Freedom to Learn: a 1985–86 report.* People for the American Way, Washington, DC.

Phillips, H. (1986). 'Pat Robertson', *Conservative Digest*, January, 84–6.

Phillips, K. P. (1982). *Post-Conservative America.* Random House, New York.

Pierard, R. V. (1985). 'Religion and the 1984 election campaign', *Rev. Rel. Res.*, Vol. 27 (2), 98–113.

Pierard, R. V. and Wright, J. L. (1984). 'No Hoosier hospitality for humanism: the Moral Majority in Indiana', in D. Bromley and A. Shupe (eds.), *New Christian Politics*, 195–212. Mercer University Press, Macon, Ga.

Pines, B. Y. (1982). *Back to Basics*. William Morrow, New York.

Popper, K. (1978). *Unended Quest: an intellectual biography*. Fontana, London.

Pound, W. and Jones, R. (1986). 'Whose victory?', *State Legislatures*, November/December, 10–22.

Pro-Family Forum (1983). *Is Humanism Molesting your Child?* Pro-Family Forum, Fort Worth.

Rabkin, J. (1986). 'The new chief, the new justice and the new court', *American Spectator*, October, 20–3.

Reavis, D. J. (1984). 'The politics of armageddon', *Texas Monthly*, Vol. 12 (October), 162–6, 235–46.

Reeves, R. (1982). *American Journey*. Simon and Schuster, New York.

Reichley, A. J. (1986). 'Religion and the future of American politics', *Pol. Sci. Qtly.*, Vol. 101, 23–47.

Rice, A. S. (1972). *The Ku Klux Klan in American Politics*. Haskell House, New York.

Richardson, J. T. (1984). 'The "old right" in action: Mormon and Catholic involvement in the equal rights amendment referendum', in D. Bromley and A. Shupe (eds.), *New Christian Politics*, 213–34. Mercer University Press, Macon, Ga.

Rifkin, J. and Howard, T. (1979). *The Emerging Order: God in the age of scarcity*. G. P. Putnam's, New York.

Robertson, R. (1983), 'Church–state relations and the world system', in T. Robbins and R. Robertson (eds.), *Church–State relations; tensions and transitions*, 5–13. Transaction, New Brunswick.

Rogers, E. M. (1962). *Diffusion of Innovation*. Free Press, New York.

Rose, T. (1983). 'Diminishing returns: the false promise of direct mail', *Washington Monthly*, June, 32–8.

Roy, R. L. (1953). *Apostles of Discord: a study of organized bigotry and disruption on the fringes of Protestantism*. Beacon Press, Boston.

Rush, G. B. (1967). 'Status inconsistency and right-wing extremism', *Amer. Sociol. Rev.*, Vol. 32, 86–92.

Safire, W. (1987). 'Elmer Gantry lives', *Washington Post*, 26 March.

Saloma, J. S. III (1984). *Ominous Politics: the new conservative labyrinth*. Hill and Wang, New York.

Schneider, J. (1986). 'Preachers and politics: new Christian right ideology and political activism among fundamentalist ministers'. Paper given at Southern Sociol. Soc., New Orleans, April.

Schuman, H., Steeh, S., and Bobo, L. (1985). *Racial Attitudes in America: trends and interpretations*. Harvard University Press, Cambridge.

Schwartz, H. (1985). *The New Right's Court Packing Campaign*. People for the American Way, Washington, DC.

Segal, D. R. (1970). 'Status inconsistency and party choice in Canada; an attempt to replicate', *Can. J. Pol. Sci.*, Vol. 3, 471–4.

Sendor, B. (1983). 'The law and religion in the public schools: a guide for the perplexed', *Popular Government*, Fall, 34–40.

—— (1984). 'The role of religion in the public school curriculum', *Popular Government*, Fall, 41–50.

Shupe, A. and Heinerman, J. (1985). 'Mormonism and the new Christian right: an emerging coalition?'. *Rev. Rel. Res*, Vol. 27 (2), 146–57.

—— and Stacey, W. (1982). *Born-again Politics: what social surveys really show*. Edwin Mellen Press, New York.

—— —— (1983). 'The Moral Majority constituency', in R. C. Liebman and R. Wuthnow (eds.), *The New Christian Right*, 104– 17. Aldine, Chicago.

Simpson, J. H. (1983). 'Moral issues and status politics', in R. C. Liebman and R. Wuthnow (eds.), *The New Christian Right*, 188–207. Aldine, Chicago.

—— (1985). 'Status inconsistency and moral issues', *J. Sci. Stud. Rel.*, Vol. 24 (2), 119–236.

Smelser, N. (1966). *The Theory of Collective Behaviour*. Routledge and Kegan Paul, London.

Smidt, C. (1983). 'Born again politics: the political behavior of evangelical Christians in the South and the non-South', in T. A. Baker, R. P. Steed, and L. W. Moreland (eds.), *Religion and Politics in the South: mass and elite perspectives*. 27–56 Praeger, New York.

Snow, D. A., Zurcher, L. A., and Ekland-Olson, S. (1980). 'Social networks and social movements: a microstructural approach to differential recruitment', *Amer. Sociol. Rev.*, Vol. 45, 787–801.

Stacey, W., Shupe, A., and Stacey, S. (1982). 'Religious values and religiosity in the textbook adoption controversy in Texas, 1981'. Paper given at Society for the Scientific Study of Religion, Providence, Rhode Island, October.

Stallings, R. A. (1973). 'Patterns of belief in social movements: clarifications from analysis of environmental groups', *Sociol. Qtly.*, Vol. 14, 465–80.

Stark, R. and Bainbridge, W. S. (1985). *The Future of Religion: secularization, revival and cult formation*. University of California Press, Berkeley

Starnes, C. E. and Singleton, R. Jun. (1977). 'Objective and subjective status inconsistency: a search for empirical correspondence', *Sociol. Qtly,*, Vol. 18, 253–66.

Stone, B. A. (1974). 'The John Birch Society: a profile', *J. Politics*, Vol. 36 (1), 184–97.

Strober, G. and Tomczak, R. (1979). *Jerry Falwell: aflame for God*. Thomas Nelson Publishers, Nashville.

Strout, C. (1974). *New Heavens and New Earth: political religion in America.* Harper and Row, New York.

Talmon, Y. (1969). 'Pursuit of the millennium: the relation between religious and social change', in N. Birnbaum and G. Lenzer (eds.), *Sociology and Religion: a book of readings*, 238–53. Basic Books, New York.

Tamney, J. B. and Johnson, S. D. (1983). 'The Moral Majority in Middletown', *J. Sci. Stud. Rel.*, Vol. 22 (2), 145–57.

—— —— (1986). 'The Christian right, morality and the law'. Unpublished paper.

Thompson, E. (1984). 'Pacs Americana: political action committees and political parties, 1980 and 1982', *Politics*, Vol. 18, 90–7.

Till, B. (1971). *The Churches' Search for Unity.* Penguin, Harmondsworth, Middx.

Turner, B. S. (1986). *Citizenship and Capitalism: the debate over reformism.* George Allen and Unwin, London.

Turner, J. (1970). *Party and Constituency: pressures on Congress.* Johns Hopkins University Press, Baltimore.

Turner, R. H. and Killian, L. M. (1965). *Collective Behavior*, Prentice-Hall, Englewood Cliffs, NJ.

Viguerie, R. A. (1981). *The New Right: we're ready to lead.* The Viguerie Co., Falls Church, Va.

Vile, M. J. C. (1976). *Politics in the USA.* Hutchinson University Library, London.

Vinz, W. L. (1972). 'The politics of Protestant fundamentalism in the 1950s and 1960s', *J. Church and State*, Vol. 14 (2), 235–60.

Wallis, R. (1979). *Salvation and Protest: studies of social and religious movements.* Frances Pinter, London.

—— and Bruce, S. (1982). 'Network and clockwork', *Sociology*, Vol. 16 (1), 102–7.

—— —— (1985). 'A comparative analysis of conservative Protestant politics', *Social Compass*, Vol. 32 (2–3), 145–61.

—— —— (1986). *Sociological Theory, Religion and Collective Action.* The Queen's University of Belfast, Belfast.

Warren, D. I. (1970). 'Status inconsistency and flying saucer sightings', *Science*, Vol. 170, 599–603.

Weber, M. (1947/1964). *The Theory of Social and Economic Organization.* Free Press, New York.

Weeks, J. (1987). 'Love in a cold climate', *Marxism Today*, January, 12–17.

Weissmann, A. (1982). 'Building a Tower of Babel', *Texas Outlook*, Winter, 10–15, 29–34.

Wilcox, C. (1986). 'Evangelicals and fundamentalists in the new Christian right: religious differences in the Ohio Moral Majority', *J. Sci. Stud. Rel.*, Vol. 25 (3), 355–63.

Wilson, B. R. (1982). *Religion in Sociological Perspective*. Oxford University Press, Oxford.

Wilson, J. (1970). *Introduction to Social Movements*. Basic Books, New York.

Wilson, K. L. and Zurcher, L. A. (1976). 'Status inconsistency and participation in social movements: an application of Goodman's hierarchical modeling', *Sociol. Qtly.*, Vol. 17, 520–33.

Wimberley, R. C. and Christenson, J. A. (1980). 'Civil religion and church and state', *Sociol. Qtly.*, Vol. 21, 35–40.

Wolfinger, R., Wolfinger, B., Prewitt, K., and Rosenhack, S. (1970). 'America's radical right: politics and ideology', in E. Dreyer and W. Rosenhack (eds.), *Political Opinion and Behaviour*, 347–82. Wadsworth, Belmont, Ca.

Wood, J. E. (1972). 'Religion and public education in historical perspective', *J. Church and State*, Vol. 14 (3), 397–414.

—— (1981). 'Tuition tax credits for nonpublic schools?', *J. Church and State*, Vol. 23 (1), 5–13.

—— (1982). " 'Scientific Creationism" and the public schools', *J. Church and State*, Vol. 24 (2), 231–43.

—— (1984). 'Religion and education in American Church–State Relations', *J. Church and State*, Vol. 26 (1), 31–54.

Yinger, J. M. and Cutler, S. J. (1984). 'The Moral Majority viewed sociologically', in D. Bromley and A. Shupe (eds.), *New Christian Politics*, 69–90. Mercer University Press, Macon, Ga.

Zald, M. N. and Ash, R. (1966). 'Social movement organizations: growth, decline and change', *Soc. Forces*, Vol. 44, 327–40.

Zanden, J. W. Vander (1960). 'The Klan revival', *Amer. J. Sociol.*, Vol. 65, 456–62.

Zurcher, L. A. and Kirkpatrick, R. G. (1976). *Citizens for decency: anti-pornography crusades as status defence*. University of Texas Press, Austin, TX.

—— Kirkpatrick, R. G., Cushing, R. G., and Bowman, C. K. (1971). 'The anti-pornography campaign: a symbolic crusade', *Social Problems*, Vol. 19 (2), 217–38.

—— and Snow, D. A. (1981). 'Collective behavior: social movements', in M. Rosenberg and R. H. Turner (eds.), *Social Psychology: selected readings*, 447–82. Basic Books, New York.

Zwier, R. (1984). 'The new Christian right and the 1980 elections', in D. Bromley and A. Shupe (eds.), *New Christian Politics*, 169–72. Mercer University Press, Macon, Ga.

Index

Southern Methodist Univ.
3 2177 01649 4542